LISREL 8:
New Statistical Features

LISREL® 8:
New Statistical Features

Karl G. Jöreskog, Dag Sörbom,

Stephen du Toit, and Mathilda du Toit

LISREL is a registered trademark, SIMPLIS and PRELIS are trademarks of Scientific Software International, Inc.

General notice: Other product names mentioned herein are used for identification purposes only and may be trademarks of their respective companies.

LISREL® 8: New Statistical Features.

Copyright © 2000 by

Karl G. Jöreskog, Dag Sörbom, Stephen du Toit, and Mathilda du Toit.

All rights reserved. Printed in the United States of America.

No part of this publication may be reproduced or distributed, or stored in a data base or retrieval system, or transmitted, in any form or by any means, without the prior written permission of the authors.

Edited by Leo Stam

Cover based on an architectural detail from Frank Lloyd Wright's Robie House.

2 3 4 5 6 7 8 9 0   02 01 00   (second printing with revisions)

Published by:

Scientific Software International, Inc.
7383 North Lincoln Avenue, Suite 100
Lincolnwood, IL 60712–1704
Tel: +1.847.675.0720
Fax: +1.847.675.2140
URL: http://www.ssicentral.com

ISBN: 0-89498-047-5

Preface

This manual is designed for reference use with all versions of LISREL: Mainframe, UNIX, MAC, and PC. This platform independency is achieved by using exclusively the 'batch mode' way of operation in the illustrations: the user writes an input (or command) file using a text editor and submits this file to the program for execution. At the present time, a fully interactive mode of working with LISREL is only available for the Windows operating system. The end of Chapter 1 has a brief description. A companion guide, called *Interactive LISREL*,[1] illustrates this interactive use in detail.

The present text builds on the three existing manuals

>   *PRELIS 2: User's Reference Guide*,
>
>   *LISREL 8: User's Reference Guide*, and
>
>   *LISREL 8: Structural Equation Modeling with the SIMPLIS Command Language*.

Those manuals introduce the reader to the methodology of structural equation modeling, both theoretically and through a variety of applications. Specifically, the *PRELIS 2: User's Reference Guide* discusses the intricacies of multivariate data screening and preparing the data for subsequent model fitting. The *LISREL 8: User's Reference Guide* elaborates on the fitting and testing of structural equation models with the use of the *LISREL* syntax, while *LISREL 8: Structural Equation Modeling with*

---

[1] Also published by Scientific Software International.

the SIMPLIS Command Language does the same with the *SIMPLIS* syntax. Chapter 1 of the current manual, *New Developments in LISREL 8* reviews the *LISREL* approach of statistical data analysis and brings the reader up-to-date through version 8 of the program.

Chapter 2, *Multilevel Modeling*, contains an extensive description of how to analyze data with a hierarchical structure. It introduces the field to the reader, presents the statistical theory, describes the file structure and syntax for this type of analysis, and takes the reader on a tutorial tour of multilevel modeling through a step-by-step description of the program in action on a variety of examples.

In Chapter 3, *Other New Statistical Features*, some powerful statistical additions to LISREL are explained: two-stage least-squares estimation, exploratory factor analysis, principal components, normal scores, and latent variable scores.

Chapter 4, *Standard Errors and Chi-Squares*, gives a detailed overview of the new standard errors and chi-squares available in *LISREL 8*. Appendix A, *Robust Standard Errors and Chi-Squares*, expands on that in a highly technical way.

Appendix B, *Why are t-Values for Error Variances Equal?*, discusses this question in some detail, while Appendix C, *Problems with Analysis of Correlation Matrices*, reviews the statistical consequences of analyzing correlations instead of covariances.

Appendix D, E, and F present a concise overview of the PRELIS, LISREL, and multilevel syntax, in a diagram format, updated with the statistical additions that are described in this manual.

This text was written by Karl Jöreskog and Dag Sörbom, except for Chapter 2, that was written by Mathilda du Toit and Stephen du Toit (both of Scientific Software International).

Chicago, March 1999.

Contents

Preface

List of examples   vii

1 New Developments in LISREL 8   1
    1.1 PRELIS 1
       1.1.1 Variables 2
       1.1.2 Data 2
       1.1.3 Matrices 2
       1.1.4 Multivariate multinomial probit regressions . . 3
       1.1.5 Asymptotic Variances and Covariances 3
       1.1.6 Simulation 4
    1.2 LISREL 4
       1.2.1 Model 5
       1.2.2 Estimation and Testing 6
             Maximum Likelihood 7
             Weighted Least Squares 7
             Weighted Least Squares based on Augmented Moment Matrix 8
             Test of the Model 9
       1.2.3 Multigroup Analysis 9
       1.2.4 General Covariance Structures 10
       1.2.5 Command Languages 10
       1.2.6 Path Diagrams 11
       1.2.7 Windows Interface 11
       1.2.8 Output 12

2 Multilevel Modeling — 13
2.1 Introduction to Multilevel Modeling — 13
2.2 Theoretical Background — 14
2.2.1 A fixed parameter linear regression model — 15
2.2.2 A level-2 model — 17
2.2.3 A general level-3 model — 19
2.2.4 Parameter estimation — 22
IGLS estimators — 27
2.2.5 Statistical inference — 27
Standard errors — 28
Contrasts — 28
Empirical Bayes estimates — 30
Likelihood ratio tests — 31
2.2.6 Multilevel logit models — 32
Introduction — 32
A level-3 logit model — 32
Estimation considerations — 37
2.3 Multilevel Input Files — 39
2.3.1 Data file — 40
2.3.2 Syntax file — 41
Syntax overview — 41
Guidelines for constructing or changing the input file — 42
OPTIONS — 42
IDn — 46
RANDOMn — 47
RESPONSE — 48
FIXED — 49
COVnPAT — 50
COVnVAL — 55
FIXVAL — 56
CONTRAST — 57
MISSING_DAT — 59
MISSING_DEP — 60
SUBPOP — 61
WEIGHT1 — 62
TITLE — 62
2.4 Multilevel Output Files — 63

		Analysis requested	63
		Data summary	64
		Fixed part of model	65
		Log-likelihood value	65
		Random part of model	66
		Random coefficient covariance and correlation matrices	67
		Technical details	68
		Empirical Bayes estimates: the output files *.BA2 and *.BA3	68
		Residuals: the output file *.RES	69
2.5	Multilevel Examples		70
	2.5.1	Analysis of 2-level repeated measures data	70
		Description of the data	71
		Variance decomposition	72
		Modeling linear growth	76
		Modeling non-linear growth	81
		Introducing a covariate while modeling non-linear growth	87
		Complex variation at level 1 of the model	89
		Conclusions	91
	2.5.2	Analysis of air traffic control data	92
		Variance decomposition model for Air Force data	93
		Non-linear model for air traffic data	96
		Including additional variables in the air traffic data analysis	99
	2.5.3	A multivariate analysis of educational data	102
		A variance decomposition model	104
		Adding explanatory variables to the model	109
	2.5.4	Analysis of CPC survey data	113
		3-level model for subset of CPC survey data	116
		Three-level model for the educational sector	121
		Three-level model for the construction sector	125

3 Other New Statistical Features — 129
3.1 Two-Stage Least-Squares — 130
3.1.1 Implementation — 132

	3.1.2	Residual Analysis	133
	3.1.3	Regression Models	134
		Example: Prediction of Grade Averages	134
	3.1.4	Econometric Models	136
		Example: Klein's Model I of US Economy	137
		Example: Tintner's Meat Market Model	141
3.2	Exploratory Factor Analysis	143
		Example: Exploratory Factor Analysis of Nine Psychological Variables	147
	3.2.1	Number of Factors	153
	3.2.2	Factor Scores	155
3.3	Principal Components	157
		Example: Principal Components of Five Meteorological Variables	158
3.4	Normal Scores	161
		Example: Normalizing the Nine Psychological Variables	162
		Example: Normalizing WORDMEAN and COUNTDOT	165
3.5	System Files	168
	3.5.1	The PRELIS System File	168
	3.5.2	The Data System File	169
	3.5.3	The Model System File	170
3.6	Latent Variable Scores	171
3.7	Interaction and Non-Linear Models	171
	3.7.1	Estimation by TSLS	172
		Example: The Kenny–Judd Model	172
	3.7.2	Estimation by Means of Latent Variable Scores	174
3.8	Simulation	175

4 Standard Errors and Chi-Squares — 179

4.1	Standard Errors	179
4.2	Chi-squares	180
4.3	LISREL implementation	181
4.4	Other Fit Statistics	182
4.5	GF File	182
4.6	Examples	183

A Robust Standard Errors and Chi-Squares — 191
A.1 Single Group: Covariance Structures — 192
Definitions — 192
Fit functions — 194
Results — 195
A.2 Single Group: Mean and Covariance Structures — 196
Fit functions — 197
A.3 Single Group: Augmented Moment Matrix — 198
Definitions — 199
Fit functions — 199
A.4 Multiple Groups: Covariance Structures — 199
Definitions — 200
Results — 201
A.5 Multiple Groups: Mean and Covariance Structures — 202
A.6 Multiple Groups: Augmented Moment Matrices — 202

B Why are t-Values for Error Variances Equal? — 203
SIMPLIS — 205
LISREL — 206

C Problems with Analysis of Correlation Matrices — 209

D PRELIS Syntax Overview — 215
D.1 PRELIS syntax diagram — 217
D.1.1 Data input commands — 217
D.1.2 Data manipulation commands — 217
Scale types — 217
Recode and label categories — 218
Transformation and creation of variables — 218
Select cases and variables — 218
D.1.3 Treatment of missing values — 219
D.1.4 Analysis and output commands — 219

E LISREL Syntax Overview — 221
E.1 LISREL syntax diagram — 223
E.1.1 Input specification commands — 223
E.1.2 General analysis specification commands — 224
E.1.3 LISREL model specification commands — 225

	E.1.4 Output specification commands	227
E.2	Notation	228
F	**Multilevel Syntax Overview**	**229**
	F.1 Multilevel syntax diagram	231

References   **233**

Author index   **241**

Subject index   **243**

List of examples
(with input and data files)

This book adds many examples to the ones that were already included with earlier versions of the program (see Jöreskog & Sörbom, 1996a–c). The examples in this book are illustrating the newly added statistical features. It is instructive to go over these examples to learn how to set up the command file for particular models and problems. We also suggest using these examples as exercises in the following ways:

- Estimate the same model with a different method of estimation
- Request other options for the output
- Formulate and test hypotheses about the parameters of the model
- Estimate a different model for the same data
- Make deliberate mistakes in the input file and see what happens

Input and data files for these examples are included with the program on the distribution media. For these files we use the following naming conventions.
Command files have the suffix LS8 for LISREL 8 files, SPL for SIMPLIS, and PR2 for PRELIS 2 files. The suffix after the period in the name of a data file refers to the type of data it contains:

- LAB for labels
- COV for covariance matrix
- COR for correlation matrix
- RAW for raw data
- DAT for a file containing several types of data

- PML for matrix of polychoric (and polyserial) correlations produced by PRELIS under listwise deletion
- KML for matrix of product-moment correlations (based on raw scores or normal scores) produced by PRELIS under listwise deletion
- ACP for asymptotic covariance matrix of the elements of a PML matrix produced by PRELIS
- ACK for asymptotic covariance matrix of the elements of a KML matrix produced by PRELIS

LIST OF EXAMPLES

Title of Example	See page
Mouse data: Variance decomposition 　Data File:　MOUSE.PSF 　Input File:　MOUSE1.PR2	70, 71, 72
Mouse data: Modeling linear growth 　Data File:　MOUSE.PSF 　Input Files:　MOUSE2.PR2, MOUSE3.PR2	70, 71, 76
Mouse data: Modeling non-linear growth 　Data File:　MOUSE.PSF 　Input File:　MOUSE4.PR2	70, 71, 81
Mouse data: Adding a covariate 　Data File:　MOUSE.PSF 　Input File:　MOUSE5.PR2	70, 71, 87
Mouse data: Complex variation 　Data File:　MOUSE.PSF 　Input File:　MOUSE6.PR2	70, 71, 89
Air traffic data: Variance decomposition 　Data File:　KANFER.PSF 　Input File:　KANFER1.PR2	92, 93
Air traffic data: Non-linear model 　Data File:　KANFER.PSF 　Input File:　KANFER2.PR2	92, 96
Air traffic data: Adding variables 　Data File:　KANFER.PSF 　Input File:　KANFER3.PR2	92, 99

LIST OF EXAMPLES

Educational data: Variance decomposition   102, 104
    Data File: JSP.PSF
    Input File: JSP1.PR2

Educational data: Adding variables   102, 109
    Data File: JSP.PSF
    Input File: JSP2.PR2

CPC data: 3-level model, all data   113, 116
    Data File: INCOME.PSF
    Input File: INCOME1.PR2, INCOME2.PR2, INCOME3.PR2

CPC data: 3-level model, education sector   113, 121
    Data File: EDUC.PSF
    Input File: EDUC.PR2

CPC data: 3-level model, construction sector   113, 125
    Data File: CONS.PSF
    Input File: CONS.PR2

Prediction of Grade Averages   134, 121
    Data File: GRAV.RAW
    Input File: GRAV.PR2

Estimating Klein's Consumption Function   137–141
    Data File: KLEIN.RAW
    Input Files: KLEIN1.PR2, KLEIN2.PR2, KLEIN3.PR2

Tintner's Meat Market Model   141
    Data File: TINTNER.COV
    Input Files: TINTNER1.SPL, TINTNER2.LS8

LIST OF EXAMPLES

Exploratory Factor Analysis of Nine Psychological Variables   147
    Data Files: NPV.RAW, NPV.KM
    Input Files: NPV1.PR2, NPV2.SPL, NPV3.LS8,
                    NPV4.SPL, NPV5.LS8, NPV6.LS8,
                    NPV7A.PR2, NPV7B.PR2, NPV7C.PR2, NPV7D.PR2

Principal Components of Five Meteorological Variables   158
    Data File: PCEX.RAW
    Input Files: PCEX1.LS8, PCEX2.LS8, PCEX3.PR2

Normalizing the Nine Psychological Variables   162
    Data File: NPV.RAW
    Input File: NPVNSC1.PR2

Normalizing WORDMEAN and COUNTDOT   165
    Data Files: NPV.RAW, WMCD.RAW
    Input Files: WMCD1.PR2, WMSCD2.PR2

The Kenny–Judd Model   172
    Data Files: KJUDD.RAW, KJUDD.PSF
    Input Files: KJTSLS1.PR2, KJTSLS2.PR2, KJUDD.PR2,
                    KENJUDD.SPL, KENJUDD.LS8, KENJUDD.PR2,

Six Psychological Variable: Pasteur & Grant–White School   183
    Data File: SPV.RAW
    Input Files: SPV.PR2, SPV1.SPL, SPV2.SPL, SPV3.SPL

Illustrating Equal T-values   205
    Data Files: EQTVAL.COV, EQTVAL.DAT
    Input Files: EQTVAL1.SPL, EQTVAL2.SPL, EQTVAL1.LS8,
                    EQTVAL2.LS8

Lawley Factor Analysis Example   211
    Data File: LAWLEY.COR
    Input Files: LAWLEY1.LS8, LAWLEY2.LS8

1 New Developments in LISREL 8

LISREL[1] consists of two programs: PRELIS and LISREL. One can run PRELIS or LISREL separately or PRELIS and LISREL in sequence. PRELIS handles everything that has to do with raw data; LISREL handles the fitting and testing of models to summary statistics produced by PRELIS. This chapter gives a short description of PRELIS 2.30 and LISREL 8.30. For further details see the three books Jöreskog & Sörbom (1996a–c) that come with the program, as well as the remainder of this book. You will find detailed instructions for every command and many real examples.

1.1 PRELIS

PRELIS is a preprocessor for LISREL. But it can also be conveniently used to provide a first descriptive look at raw data even when no LISREL analysis is intended or when further analysis will be done by other programs. PRELIS gives a fast and efficient data screening and data summarization by analyzing all variables jointly (as opposed to one variable at a time). There is no limit on sample size.

Experience with consultation with users on LISREL has shown that they are often not sufficiently familiar with characteristics and problems in the raw data when they set out to estimate and test a LISREL model. Problems in the raw data can often account for peculiarities that occur when estimating and testing LISREL models and may, in fact, invalidate the whole LISREL analysis.

---

[1]This chapter was written by Karl Jöreskog

1.1.1 Variables

A fundamental principle in PRELIS is the distinction between variables of different scale types. PRELIS 2 distinguishes between the following types of variables: continuous (normal or non-normal), ordinal (up to 15 categories), censored (above, below and both), and fixed or random covariates. PRELIS gives univariate and multivariate tests of normality for continuous variables. Ordinal, censored, and fixed variables require quite different treatment than continuous variables. The unique handling of ordinal variables by PRELIS includes estimation of thresholds, test of equal thresholds, and test of underlying normality, as well as estimation of polychoric correlations under unconstrained, fixed, or equal thresholds. These alternative procedures are particularly important when the same ordinal variables are used at several occasions (longitudinal studies) or in several groups (multi-group studies).

1.1.2 Data

The data may be cross-sectional, time series, longitudinal, single samples, multiple samples, and case-weighted. Missing values in the raw data may be handled by pairwise deletion, listwise deletion, and imputation based on matching. Variables may be transformed by various transformations: linear, normal scores, power, and logarithmic, and new variables can be formed as functions of already defined variables. Both ordinal and continuous variables can be recoded. Variables may be selected for inclusion or exclusion and cases may be selected for inclusion based on different criteria for different variables. Files containing data on different variables on the same cases can be merged into one file.

1.1.3 Matrices

For the variables and cases remaining after transformations and or case selections, PRELIS can compute many different types of matrices: covariance matrix, moment matrix, augmented moment matrix, and different kinds of correlation matrices: product moment (Pearson) correlations, canonical correlations, polychoric and polyserial correlations, Spearman rank correlations, Kendall's tau-c correlations.

The mean vector and any one of the above matrices may be saved in files to be read by LISREL.

1.1.4 Multivariate multinomial probit regressions

A major new feature in PRELIS 2 is the distinction between y- and x-variables. The x-variables can be fixed or random variables (covariates). If they are random, their joint distribution is unspecified and assumed not to contain any parameters of interest. These x-variables can be dummy-coded categorical variables or measured variables on an interval scale assumed not to contain measurement error. The y-variables can be continuous, censored, or ordinal variables. PRELIS 2 will estimate the conditional covariance matrix of y for given x and the unconditional joint covariance matrix of y and x and its asymptotic covariance matrix. These can be used with WLS in LISREL 8 in an analysis with fixed-x.

1.1.5 Asymptotic Variances and Covariances

PRELIS can produce consistent estimates of the asymptotic covariance matrix of the elements of the following matrices:

(i) covariance matrix of continuous variables,
(ii) correlation matrix of product moment correlations for continuous variables,
(iii) correlation matrix of polychoric correlations for ordinal variables.

For all other cases, estimates of the asymptotic covariance matrix can be obtained by bootstrapping.

The asymptotic covariance matrix is saved in binary form for use with weighted least squares (WLS, equivalent to ADF) in LISREL. Alternatively, one may choose to save only the asymptotic variances for use with DWLS in LISREL.

1.1.6 Simulation

By simulation we mean here drawing random samples of data from some population, estimating various parameters from each sample for the purpose of studying the mean and variance and other characteristics of the distribution of these parameter estimates. There are two techniques for drawing the random samples: bootstrap and Monte Carlo. In bootstrapping the random samples are drawn from an original sample, which usually is a sample of empirical data but which can also be a set of artificial data. In Monte Carlo sampling, the samples are generated from randomly generated variables so no real data is involved. Various combinations of the two techniques are also possible.

In Monte Carlo experiments, raw data can be generated directly with PRELIS, and covariance or correlation matrices can be computed directly *without saving or storing the raw data*. There are almost unlimited possibilities of generating normal and non-normal variables with specified properties. One can also generate discrete variables (ordinal or categorical).

1.2 LISREL

The LISREL model is a formal mathematical model which has to be given substantive content in each application. The meaning of the terms in the model varies from one application to another. The formal LISREL model defines a large class of models within which one can work and this class contains several useful subclasses as special cases.

In its most general form, the LISREL model consists of a set of linear structural equations. Variables in the equation system may be either directly observed variables or unmeasured latent (theoretical) variables that are not observed but relate to observed variables. The model assumes that there is a "causal" structure among a set of latent variables, and that the observed variables are indicators or symptoms of the latent variables. Sometimes the latent variables appear as linear composites of observed variables, other times as intervening variables in a "causal chain." The

1.2 LISREL

LISREL methodology is particularly designed to accommodate models that include latent variables, measurement errors, and reciprocal causation.

In addition, LISREL covers a wide range of models useful in the social and behavioral sciences, including confirmatory factor analysis, path analysis, econometric models for time series data, recursive and non-recursive models for cross-sectional and longitudinal data, and covariance structure models.

1.2.1 Model

In its most general form the LISREL model is defined as follows. Consider random vectors $\boldsymbol{\eta}' = (\eta_1, \eta_2, \ldots, \eta_m)$ and $\boldsymbol{\xi}' = (\xi_1, \xi_2, \ldots, \xi_n)$ of latent dependent and independent variables, respectively, and the following system of linear structural relations

$$\boldsymbol{\eta} = \boldsymbol{\alpha} + \mathbf{B}\boldsymbol{\eta} + \boldsymbol{\Gamma}\boldsymbol{\xi} + \boldsymbol{\zeta}, \tag{1.1}$$

where $\boldsymbol{\alpha}$ is a vector of interecept terms, $\mathbf{B}$ and $\boldsymbol{\Gamma}$ are coefficient matrices and $\boldsymbol{\zeta}' = (\zeta_1, \zeta_2, \ldots, \zeta_m)$ is a random vector of residuals (errors in equations, random disturbance terms). The elements of $\mathbf{B}$ represent direct effects of η-variables on other η-variables and the elements of $\boldsymbol{\Gamma}$ represent direct effects of ξ-variables on η-variables. It is assumed that $\boldsymbol{\zeta}$ is uncorrelated with $\boldsymbol{\xi}$ and that $\mathbf{I} - \mathbf{B}$ is non-singular.

Vectors $\boldsymbol{\eta}$ and $\boldsymbol{\xi}$ are not observed, but instead vectors $\mathbf{y}' = (y_1, y_2, \ldots, y_p)$ and $\mathbf{x}' = (x_1, x_2, \ldots, x_q)$ are observed, such that

$$\mathbf{y} = \boldsymbol{\tau}_y + \boldsymbol{\Lambda}_y \boldsymbol{\eta} + \boldsymbol{\epsilon}, \tag{1.2}$$

and

$$\mathbf{x} = \boldsymbol{\tau}_x + \boldsymbol{\Lambda}_x \boldsymbol{\xi} + \boldsymbol{\delta}, \tag{1.3}$$

where $\boldsymbol{\epsilon}$ and $\boldsymbol{\delta}$ are vectors of error terms (errors of measurement or measure-specific components) assumed to be uncorrelated with $\boldsymbol{\eta}$ and $\boldsymbol{\xi}$, respectively. These equations represent the multivariate regressions of $\mathbf{y}$ on

η and of $\mathbf{x}$ on $\boldsymbol{\xi}$, respectively, with $\boldsymbol{\Lambda}_y$ and $\boldsymbol{\Lambda}_x$ as regression matrices and $\boldsymbol{\tau}_y$ and $\boldsymbol{\tau}_x$ as vectors of constant intercept terms. It is convenient to refer to $\mathbf{y}$ and $\mathbf{x}$ as the observed variables and η and $\boldsymbol{\xi}$ as the latent variables.

Let $\boldsymbol{\kappa}$ be the mean vector of $\boldsymbol{\xi}$. $\boldsymbol{\Phi}$ and $\boldsymbol{\Psi}$ the covariance matrices of $\boldsymbol{\xi}$ and $\boldsymbol{\zeta}$, $\boldsymbol{\Theta}_\epsilon$ and $\boldsymbol{\Theta}_\delta$ the covariance matrices of ϵ and δ, and $\boldsymbol{\Theta}_{\delta\epsilon}$ the covariance matrix between δ and ϵ. Then it follows that the mean vector $\boldsymbol{\mu}$ and covariance matrix $\boldsymbol{\Sigma}$ of $\mathbf{z} = (\mathbf{y}', \mathbf{x}')'$ are

$$\boldsymbol{\mu} = \begin{pmatrix} \boldsymbol{\tau}_y + \boldsymbol{\Lambda}_y(\mathbf{I} - \mathbf{B})^{-1}(\boldsymbol{\alpha} + \boldsymbol{\Gamma}\boldsymbol{\kappa}) \\ \boldsymbol{\tau}_x + \boldsymbol{\Lambda}_x \boldsymbol{\kappa} \end{pmatrix}, \qquad (1.4)$$

$$\boldsymbol{\Sigma} = \begin{pmatrix} \boldsymbol{\Lambda}_y \mathbf{A}(\boldsymbol{\Gamma}\boldsymbol{\Phi}\boldsymbol{\Gamma}' + \boldsymbol{\Psi})\mathbf{A}'\boldsymbol{\Lambda}_y' + \boldsymbol{\Theta}_\epsilon & \boldsymbol{\Lambda}_y \mathbf{A}\boldsymbol{\Gamma}\boldsymbol{\Phi}\boldsymbol{\Lambda}_x' + \boldsymbol{\Theta}_{\delta\epsilon}' \\ \boldsymbol{\Lambda}_x \boldsymbol{\Phi}\boldsymbol{\Gamma}'\mathbf{A}'\boldsymbol{\Lambda}_y' + \boldsymbol{\Theta}_{\delta\epsilon} & \boldsymbol{\Lambda}_x \boldsymbol{\Phi}\boldsymbol{\Lambda}_x' + \boldsymbol{\Theta}_\delta \end{pmatrix}, \qquad (1.5)$$

where $\mathbf{A} = (\mathbf{I} - \mathbf{B})^{-1}$.

The elements of $\boldsymbol{\mu}$ and $\boldsymbol{\Sigma}$ are functions of the elements of $\boldsymbol{\kappa}$, $\boldsymbol{\alpha}$, $\boldsymbol{\tau}_y$, $\boldsymbol{\tau}_x$, $\boldsymbol{\Lambda}_y$, $\boldsymbol{\Lambda}_x$, $\mathbf{B}$, $\boldsymbol{\Gamma}$, $\boldsymbol{\Phi}$, $\boldsymbol{\Psi}$, $\boldsymbol{\Theta}_\epsilon$, $\boldsymbol{\Theta}_\delta$, and $\boldsymbol{\Theta}_{\delta\epsilon}$ which are of three kinds:

- *fixed parameters* that have been assigned specified values,
- *constrained parameters* that are unknown but linear or non-linear functions of one or more other parameters, and
- *free parameters* that are unknown and not constrained.

1.2.2 Estimation and Testing

A LISREL model may be estimated by seven different methods: instrumental variables (IV), two-stage least squares (TSLS), unweighted least squares (ULS), generalized least squares (GLS), maximum likelihood (ML), weighted least squares (WLS), and diagonally weighted least squares (DWLS). Under general assumptions, all methods give consistent estimates of parameters. TSLS and IV are non-iterative and very fast. They are used to compute starting values for the other methods but can also be requested as final estimates. ULS, GLS, ML, WLS, and DWLS estimates are obtained

1.2 LISREL

by an iterative procedure that minimizes a particular fit function. WLS requires an estimate of the asymptotic covariance matrix of the sample variances and covariances or correlations being analyzed. Similarly, DWLS requires an estimate of the asymptotic variances of the sample variances and covariances or correlations being analyzed. These asymptotic variances and covariances are obtained by PRELIS which saves them in a file to be read by LISREL.

Let θ be a vector of all free and independent parameters of the model. Then the mean vector μ and covariance matrix Σ of z are functions of θ. Let $z_1, z_2, \ldots, z_N$ be N independent observations of the vector z and let $\bar{z}$ and S be the sample mean vector and covariance matrix. Parameter estimates are obtained by minimizing some fit function

$$\mathsf{F}(\theta) = \mathsf{F}(\bar{z}, S, \mu(\theta), \Sigma(\theta))$$

of $\mu(\theta)$ and $\Sigma(\theta)$, of which only the most important are considered here.

Maximum Likelihood

The maximum likelihood (ML) approach will estimate θ by minimizing the fit function

$$\mathsf{F}(\theta) = \log||\Sigma|| + \mathrm{tr}(S\Sigma^{-1}) - \log||S|| - k + (\bar{z} - \mu)'\Sigma^{-1}(\bar{z} - \mu) \ . \quad (1.6)$$

where k is the number of variables in z.

This fit function assumes that the observed variables z have a multinormal distribution.

Weighted Least Squares

The weighted least squares (WLS) approach will estimate θ by minimizing the fit function

$$\mathsf{F}(\theta) = (s - \sigma)'W^{-1}(s - \sigma) + (\bar{z} - \mu)'S^{-1}(\bar{z} - \mu) \ , \quad (1.7)$$

where $\mathbf{s}' = (s_{11}, s_{21}, s_{22}, s_{31}, \ldots, s_{kk})$, $\boldsymbol{\sigma}' = (\sigma_{11}, \sigma_{21}, \sigma_{22}, \sigma_{31}, \ldots, \sigma_{kk})$, and $\mathbf{W}$ is a symmetric positive definite matrix. The usual way of choosing $\mathbf{W}$ in weighted least squares is to let $\mathbf{W}$ be a consistent estimate of the asymptotic covariance matrix of $\mathbf{s}$.

WLS does not assume multivariate normality of $\mathbf{z}$ but it does assume that $\bar{\mathbf{z}}$ and $\mathbf{S}$ are asymptotically independent.

Weighted Least Squares based on Augmented Moment Matrix

To avoid the mentioned problems associated with ML and WLS, one can use the augmented moment matrix

$$\mathbf{A} = (1/N) \sum_{c=1}^{N} \begin{pmatrix} \mathbf{z}_c \\ 1 \end{pmatrix} \begin{pmatrix} \mathbf{z}'_c & 1 \end{pmatrix} = \begin{pmatrix} \mathbf{S} + \bar{\mathbf{z}}\bar{\mathbf{z}}' & \bar{\mathbf{z}} \\ \bar{\mathbf{z}}' & 1 \end{pmatrix}. \quad (1.8)$$

This is the matrix of sample moments about zero for the vector $\mathbf{z}$ augmented with a variable which is constant equal to one for every case. The corresponding population matrix is

$$(\alpha_{ij}) = \mathsf{E} \begin{pmatrix} \mathbf{z} \\ 1 \end{pmatrix} \begin{pmatrix} \mathbf{z}' & 1 \end{pmatrix} = \begin{pmatrix} \boldsymbol{\Sigma} + \boldsymbol{\mu}\boldsymbol{\mu}' & \boldsymbol{\mu} \\ \boldsymbol{\mu}' & 1 \end{pmatrix}. \quad (1.9)$$

Note that the last element in these matrices is a fixed constant equal to 1. Let

$$\begin{aligned} \mathbf{a}' &= (a_{11}, a_{21}, a_{22}, a_{31}, \ldots, a_{k+1,k}, 1) \\ \boldsymbol{\alpha}' &= (\alpha_{11}, \alpha_{21}, \alpha_{22}, \alpha_{31}, \ldots, \alpha_{k+1,k}, 1). \end{aligned}$$

Then another weighted least squares fit function (WLSA) is:

$$F(\boldsymbol{\theta}) = (\mathbf{a} - \boldsymbol{\alpha})' \mathbf{W}_a^- (\mathbf{a} - \boldsymbol{\alpha}), \quad (1.10)$$

where $\mathbf{W}_a$ is a consistent estimate of the covariance matrix of $\mathbf{a}$ and $\mathbf{W}_a^-$ is a Moore-Penrose generalized inverse of $\mathbf{W}_a$. Note that since the last element in $\mathbf{a}$ is a fixed constant, the last row of $\mathbf{W}_a$ is zero. Hence, $\mathbf{W}_a$ is singular.

Test of the Model

To test the model, one may use $c = N - 1$ times the minimum value of the fit function. If the model holds and is identified, c is approximately distributed in large samples as χ^2 with $d = s - t$ degrees of freedom, where $s = k(k+1)/2$ and t is the number of independent parameters estimated. If the model holds only approximately, c is distributed as non-central χ^2 with $s - t$ degrees of freedom and non-centrality parameter λ that may be estimated as $\hat{\lambda} = \mathsf{Max}\{(c-d), 0\}$ One can also set up a confidence interval for λ.

Once the validity of the model has been established, one can test structural hypotheses about the parameters $\boldsymbol{\theta}$ in the model such that

- certain θ's have particular values (fixed parameters)
- certain θ's are equal (equality constraints)
- certain θ's are specified linear or nonlinear functions of of other parameters.

1.2.3 Multigroup Analysis

LISREL can also analyze data from several groups or populations. These may be different nations, states, or regions, culturally or socioeconomically different groups, groups of individuals selected on the basis of some known or unknown selection variables, groups receiving different treatments, and control groups, etc. In fact, they may be any set of mutually exclusive groups of individuals that are clearly defined. It is assumed that a number of variables have been measured on a number of individuals from each population. This approach is particularly useful in comparing a number of treatment and control groups regardless of whether individuals have been assigned to the groups randomly or not. Any LISREL model may be specified and fitted for each group of data.

Let $\bar{\mathbf{z}}_g$ and $\mathbf{S}_g$ be the sample mean vector and covariance matrix in group g, and let $\boldsymbol{\mu}_g(\boldsymbol{\theta})$ and $\boldsymbol{\Sigma}_g(\boldsymbol{\theta})$ be the corresponding population mean vector and covariance matrix $g = 1, 2, \ldots, G$. The fit function for the multigroup case is defined as

$$F(\boldsymbol{\theta}) = \sum_{g=1}^{G} \frac{N_g}{N} F_g(\boldsymbol{\theta}) , \qquad (1.11)$$

where $F_g(\boldsymbol{\theta}) = F(\bar{\mathbf{z}}_g, \mathbf{S}_g, \boldsymbol{\mu}_g(\boldsymbol{\theta}), \boldsymbol{\Sigma}_g(\boldsymbol{\theta}))$ is any of the fit functions defined for a single group. Here N_g is the sample size in group g and $N = N_1 + N_2 + \cdots + N_G$ is the total sample size. To test the model, one can again use $c = (N-1)$ times the minimum of F as a χ^2 with degrees of freedom $d = Gk(k+1)/2 - t$.

1.2.4 General Covariance Structures

One can specify general covariance structures directly, even such structures that cannot be obtained by a LISREL model. One can also specify additional parameters which are not in the model but such that parameters in the model are linear or non-linear functions of these.

1.2.5 Command Languages

LISREL 8 accepts two different command languages in the input file called LISREL and SIMPLIS.

In the LISREL command language, the model is formulated in matrix form using Greek matrix notation and all commands, keywords, and options are in two-character words.

The SIMPLIS command language is in plain English. All that is required is to name all observed and latent (if any) variables and to formulate the model to be estimated. The model can be specified either as paths or as relationships (equations) or as a path diagram at run time. It is not necessary to be familiar with the LISREL model or any of its submodels. No Greek or matrix notations are required. There are no complicated options to learn. Anyone who can formulate the model as a path diagram can use the SIMPLIS command language immediately.

One can use either the SIMPLIS language or the LISREL syntax in the input file but the two languages cannot be mixed in the same input file.

Beginning users of LISREL and users who often make mistakes when they specify the LISREL model will benefit greatly from using the SIMPLIS language, as this is much easier to learn and reduces the possibilities for mistakes to a minimum. Experienced users who seldom or never make mistakes in the model specification, may want to continue to use the LISREL command language.

1.2.6 Path Diagrams

High quality path diagrams of the models can be displayed on screen and printed.[2] The path diagrams contain parameter estimates and t-values, and for fixed and constrained parameters, modification indices and expected (predicted) parameter values. One can also obtain path diagrams for the conceptual model, *i.e.*, without any estimates, and for the completely standardized solution. Path diagrams can be saved in Windows metafile format for inclusion in other documents.

Models can be changed interactively by adding or deleting paths and/or specifying equality constraints directly on the screen. Fit statistics can be viewed at the same time as a path diagram. In this interactive mode of working, several models can be studied without changing the original input command file.

1.2.7 Windows Interface

With version 8.20 there is a new Windows interface which allows for three different modes of input:

- Command files in either SIMPLIS or LISREL command language.
- Dialog box interface for either SIMPLIS or LISREL command language.
- Path diagram can be drawn directly from scratch to define the model. Path diagrams produced by the program can be revised by drawing to satisfy the user's taste.

---

[2]This feature is only available in the Windows version of LISREL.

One can switch between all three modes of input. For example, a command file fills in information in the dialog boxes, and information filled in via dialog boxes produces lines in the command file.

Whether SIMPLIS or LISREL language is used, there are four options for the output. Output may be obtained in ASCII (plain text) form, in RTF (Rich Text Format) form, in HTML (HyperText Markup Language) form, or in LaTeX form. RTF output can be imported into all major word processing software without loosing formatting information. HTML output can be used in Internet browsers. Output in LaTeX form can easily be edited and incorporated into LaTeX documents.

The Windows interface has been much improved in version 8.30. The complete Windows interface for LISREL 8 is described in detail *Interactive LISREL*.[3]

1.2.8 Output

The output from LISREL gives many quantities:

- Parameter estimates, standard errors, and t-values
- Residuals and standardized residuals
- Stemleaf plots of residuals and standardized residuals
- Q-plot of standardized residuals
- Modification indices and estimated parameter change
- Parameter plot of concentrated fit function
- 31 different measures of fit
- Test of both exact and close fit
- Two different kinds of standardized solutions, one in which only the latent variables are standardized and one in which both latent and observed variables are standardized
- Total and indirect effects and their standard errors
- Standardized total and indirect effects
- Test of admissibility of model
- Test of identification of model

---

[3] Also published by SSI.

2 Multilevel Modeling

This chapter[1] presents and illustrates the newly added multilevel modeling features in LISREL 8.30.

It starts out with a brief and general introduction, followed by an explanation of the statistical theory behind multilevel analysis.

The remainder of the chapter is devoted to how to do actual multilevel analysis. First, through an overview of the syntax for creating an input file. Next, through a step-by-step discussion of several examples.

2.1 Introduction to Multilevel Modeling

The analysis of data with a hierarchical structure has been described in the literature under various names. It is known as hierarchical modeling, random coefficient modeling, latent curve modeling, growth curve modeling or multilevel modeling. The basic underlying structure of measurements nested within units at a higher level of the hierarchy is, however, common to all. In a growth model, with repeated measures, for example, the measurements or outcomes are nested within the experimental units (second level units) of the hierarchy. We can describe these outcomes as a sum of effects for the individual measurement and for the experimental unit for which the measurement was made. Regression coefficients may be present at some or all of the levels, and variance components at different levels of the hierarchy may also be obtained.

---

[1]Written by Matilda du Toit and Stephen du Toit, Scientific Software International, Chicago

Inference can be drawn from available data for such a model for the population means at any level. Hierarchical models are particularly useful in the modeling of data from complex surveys, as cluster or multi-stage sample designs are frequently used for populations with a hierarchical structure. Ignoring the hierarchical structure of data can have serious implications, as the use of alternatives such as aggregation and disaggregation of information to another level can induce high collinearity among predictors and large or biased standard errors for the estimates. Standard fixed parameter regression models do not allow for the exploration of variation between groups, which may be of interest in its own right. For a discussion of the effects of these alternatives, see Bryk & Raudenbush (1992), Longford (1987), and Rasbash (1993).

In contrast, multilevel or hierarchical modeling provides the opportunity to study variation at different levels of the hierarchy. Such a model can also include separate regression coefficients at different levels of the hierarchy that have no meaning without recognition of the hierarchical structure of the population. The dependence of repeated measurements belonging to one experimental unit in a typical growth curve analysis, for example, is taken into account with this approach. In addition, the data to be analyzed need not be balanced in nature. This has the advantage that estimates can also be obtained for units for which a very limited amount of information is available.

In the examples given here, we primarily focus on the application of this approach to growth curve models and repeated measurements data. To illustrate the wide applicability of this analysis tool, we also include examples of the analysis of data of an educational nature and data from the 1995 CPC survey.

2.2 Theoretical Background

In this section the concept of multilevel modeling is introduced. A fixed parameter linear regression model is considered first (p. 15), followed by a level-2 model (p. 17), and finally a general level-3 model (p. 19).

Next, a brief overview of estimation procedures that may be used for the analysis of unbalanced hierarchical data is given (p. 22). Then, statistical inference is discussed, starting on p. 27.

Survey data in the social sciences are usually of a categorical nature. In Section 2.2.6 the analysis of data with categorical response variables is considered.

2.2.1 A fixed parameter linear regression model

Consider the dental measurement data set first analyzed by Potthoff & Roy (1964). The data set contains the dental measurements of 11 girls and 16 boys at ages 8, 10, 12, and 14 years. Each measurement is the distance in millimeters between the center of the pituitary and the pterygomaxillary fissure.

Suppose, we wish to investigate the relationship between the measurements $y_{ij}, j = 1, 2, 3, 4$ for child i and the ages at which the measurement were taken. Denote these ages for individual i by x_{i1}, x_{i2}, x_{i3}, and x_{i4}.

Traditionally, a single linear equation may be estimated by pooling all 27 cases. The measurement may then be expressed as a linear function of the ages at which measurements are taken and could be written as

$$y_{ij} = \beta_0 + \beta_1 x_{ij} + e_{ij}, \qquad j = 1, 2, 3, 4. \qquad (2.1)$$

Let

$$\mathbf{x}'_{ij} = [\ 1 \quad x_{ij}\ ]$$

and

$$\boldsymbol{\beta}' = [\ \beta_0 \quad \beta_1\ ]$$

From (2.1), the measurement for individual i can be rewritten as

$$y_{ij} = \mathbf{x}'_{ij}\boldsymbol{\beta} + \mathbf{e}_{ij}, \qquad i = 1, 2, \ldots, N\ ;\ j = 1, 2, 3, 4\ .$$

where $N = 27$ denotes the total number of children for which measurements were available.

Using matrix notation, the set of regression equations given above may be written as

$$\mathbf{y} = \mathbf{X}\boldsymbol{\beta} + \mathbf{e},$$

where

$$\mathbf{y} = \begin{bmatrix} \mathbf{y}_1 \\ \vdots \\ \mathbf{y}_N \end{bmatrix}, \quad \mathbf{X} = \begin{bmatrix} \mathbf{X}_1 \\ \vdots \\ \mathbf{X}_N \end{bmatrix}, \quad \text{and} \quad \mathbf{e} = \begin{bmatrix} \mathbf{e}_1 \\ \vdots \\ \mathbf{e}_N \end{bmatrix}.$$

It is assumed that $e_{11}, e_{12}, \ldots$ are uncorrelated with mean zero and constant variance σ^2. Thus,

$$\mathsf{E}(\mathbf{e}) = \mathbf{0} \tag{2.2}$$

and

$$\mathsf{Cov}(\mathbf{e}, \mathbf{e}') = \sigma^2 \mathbf{I}. \tag{2.3}$$

Under the assumptions given by (2.2) and (2.3), the ordinary least squares estimator $\hat{\boldsymbol{\beta}}$ of $\boldsymbol{\beta}$ is obtained as

$$\hat{\boldsymbol{\beta}} = (\mathbf{X}'\mathbf{X})^{-1}\mathbf{X}'\mathbf{y}$$

where

$$\mathsf{E}(\hat{\boldsymbol{\beta}}) = \boldsymbol{\beta}$$

and

$$\mathsf{Cov}(\hat{\boldsymbol{\beta}}, \hat{\boldsymbol{\beta}}') = \sigma^2 (\mathbf{X}'\mathbf{X})^{-1}.$$

2.2.2 A level-2 model

In this case the individual children are the level-2 units and the dental measurements at different ages the level-1 units. There are four level-1 units nested within each level-2 unit and there are 27 level-2 units.

The fact that the regression coefficients usually vary from one individual to another, may be accommodated by regarding the unknown regression parameters as random variables with mean β and covariance matrix Φ.

The model for the 27 level-2 units can then be defined as

$$\mathbf{y}_i = \mathbf{X}\mathbf{b}_i + \mathbf{e}_i , \quad i = 1, 2, \ldots, 27 \qquad (2.4)$$

where the 4×2 matrix $\mathbf{X}$ is given by

$$\mathbf{X} = \begin{bmatrix} 1 & 8 \\ 1 & 10 \\ 1 & 12 \\ 1 & 14 \end{bmatrix}$$

with the first column denoting the intercept term and the second column giving the ages at which measurements were made.

It is assumed that $\mathbf{b}_1, \mathbf{b}_2, \ldots, \mathbf{b}_{27}$ are a random sample from a multivariate normal distribution with

$$\mathsf{E}(\mathbf{b}_i) = \beta$$

and

$$\mathsf{Cov}(\mathbf{b}_i, \mathbf{b}_i') = \Phi .$$

The vectors $\mathbf{e}_1, \mathbf{e}_2, \ldots, \mathbf{e}_{27}$ are assumed to be independently and identically distributed as $N(\mathbf{0}, \sigma^2 \mathbf{I})$ independent of $\mathbf{b}_1, \mathbf{b}_2, \ldots, \mathbf{b}_{27}$. Under these assumptions it follows that

$$\mathsf{E}(\mathbf{y}_i) = \mathbf{X}\beta \qquad (2.5)$$

and

$$\text{Cov}(\mathbf{y}_i, \mathbf{y}_i') = \mathbf{X}\boldsymbol{\Phi}\mathbf{X}' + \sigma^2\mathbf{I} . \qquad (2.6)$$

If, however, the gender of an individual is to be taken into account, (2.4) can be rewritten as

$$\mathbf{y}_i = \mathbf{X}\mathbf{b}_i + \mathbf{e}_i , \qquad (2.7)$$

where

$$\mathbf{b}_i = \begin{bmatrix} \beta_0 + u_{i1} \\ \beta_1 + u_{i2} \\ \beta_2 + 0 \end{bmatrix}$$

and where β_2 denotes the gender coefficient.

It is more convenient to write (2.7) as

$$\mathbf{y}_i = \mathbf{X}_{(f)}\boldsymbol{\beta} + \mathbf{X}_{(2)}\mathbf{u}_i + \mathbf{e}_i , \qquad i = 1, 2, \ldots, 27 . \qquad (2.8)$$

The matrix $\mathbf{X}_{(f)}$ is the design matrix for the fixed part of the model. If gender is coded '1' for boys and '-1' for girls, the matrix $\mathbf{X}_{(f)}$, in the case of a female, is given by

$$\mathbf{X}_{(f)} = \begin{bmatrix} 1 & 8 & -1 \\ 1 & 10 & -1 \\ 1 & 12 & -1 \\ 1 & 14 & -1 \end{bmatrix} .$$

The vector $\mathbf{b}_i$ defines a model with intercept and slope coefficients which are allowed to vary over the units. A fixed gender effect is also included. The vector $\boldsymbol{\beta}$ as given in (2.8) contains coefficients for the fixed part of the model, while the vector $\mathbf{u}_i$ contains those coefficients allowed to vary over level-2 units.

2.2 THEORETICAL BACKGROUND

The matrix $\mathbf{X}_{(2)}$ is the random parameter matrix on level 2 of the model and is given by

$$\mathbf{X}_{(2)} = \begin{bmatrix} 1 & 8 \\ 1 & 10 \\ 1 & 12 \\ 1 & 14 \end{bmatrix}.$$

Let

$$\mathsf{E}(\mathbf{u}_i) = \mathbf{0}$$

and

$$\mathsf{Cov}(\mathbf{u}_i, \mathbf{u}'_i) = \boldsymbol{\Phi}_{(2)}$$

while $\mathbf{e}_1, \mathbf{e}_2, \ldots, \mathbf{e}_{27}$ are assumed to be identically and independently distributed as $N(\mathbf{0}, \sigma^2 \mathbf{I})$. Then (*cf.* (2.5) and (2.6)),

$$\mathsf{E}(\mathbf{y}_i) = \mathbf{X}_{(f)}\boldsymbol{\beta} \tag{2.9}$$

and

$$\mathsf{Cov}(\mathbf{y}_i, \mathbf{y}'_i) = \mathbf{X}_{(2)} \boldsymbol{\Phi}_{(2)} \mathbf{X}'_{(2)} + \sigma^2 \mathbf{I}. \tag{2.10}$$

2.2.3 A general level-3 model

Consider the situation where a response variable y may depend on a set of p predictors $x_1, x_2, \ldots, x_p$. The general level-3 model is defined as

$$y_{ijk} = \mathbf{x}'_{(f)ijk}\boldsymbol{\beta} + \mathbf{x}'_{(3)ijk}\mathbf{v}_i + \mathbf{x}'_{(2)ijk}\mathbf{u}_i + \mathbf{x}'_{(1)ijk}e_{ijk}, \tag{2.11}$$

where

$i = 1, 2, \ldots, N$ denotes level-3 units (*e.g.*, educational departments),

$j = 1, 2, \ldots, n_i$ denotes level-2 units (*e.g.*, schools), and

$k = 1, 2, \ldots, n_{ij}$ denotes level-1 units (*e.g.*, pupils).

$\mathbf{x}'_{(f)ijk} : 1 \times s$ is a typical row of the design matrix of the fixed part of the model, the elements being a subset of the p predictors.

$\mathbf{x}'_{(3)ijk} : 1 \times q$ is a typical row of the design matrix for the random part at level 3, the elements being a subset of the p predictors.

$\mathbf{x}'_{(2)ijk} : 1 \times m$ is a typical row of the design matrix for the random part at level 2, the elements being a subset of the p predictors.

$\mathbf{x}'_{(1)ijk} : 1 \times r$ is a typical row of the design matrix for the random part at level 1, the elements being a subset of the p predictors.

$\boldsymbol{\beta} : s \times 1$ is a vector of fixed, but unknown parameters to be estimated.

It is assumed that $\mathbf{v}_1, \mathbf{v}_2, \ldots, \mathbf{v}_N$ are independently and identically distributed with mean $\mathbf{0}$ and covariance matrix $\boldsymbol{\Phi}_{(3)}$. It is further assumed that $\mathbf{u}_{i1}, \mathbf{u}_{i2}, \ldots, \mathbf{u}_{in_i}$ are *i.i.d.* with mean $\mathbf{0}$ and covariance matrix $\boldsymbol{\Phi}_{(2)}$, while $\mathbf{e}_{ij1}, \mathbf{e}_{ij2}, \ldots, \mathbf{e}_{ijn_{ij}}$ are *i.i.d.* with mean $\mathbf{0}$ and covariance matrix $\boldsymbol{\Phi}_{(1)}$. Finally, it is assumed that $\mathbf{v}_i$, $\mathbf{u}_{ij}$, and $\mathbf{e}_{ijk}$ are independent.

Let

$$\mathbf{y}_i = \begin{bmatrix} \mathbf{y}_{i1} \\ \mathbf{y}_{i2} \\ \vdots \\ \mathbf{y}_{in_i} \end{bmatrix} \qquad (2.12)$$

where $\mathbf{y}_{ij}$ denotes the $n_{ij} \times 1$ vector of responses for the i-th level-3 unit and the j-th level-2 unit. Note that $\mathbf{y}_{ij}$ can be expressed as

2.2 THEORETICAL BACKGROUND

$$\mathbf{y}_{ij} = \mathbf{X}_{(f)ij}\boldsymbol{\beta} + \mathbf{X}_{(3)ij}\mathbf{v}_i + \mathbf{X}_{(2)ij}\mathbf{u}_{ij} + \begin{bmatrix} \mathbf{x}'_{(1)ij1}\mathbf{e}_{ij1} \\ \mathbf{x}'_{(1)ij2}\mathbf{e}_{ij2} \\ \vdots \\ \mathbf{x}'_{(1)ijn_{ij}}\mathbf{e}_{ijn_{ij}} \end{bmatrix}, \quad (2.13)$$

where

$$\mathbf{X}_{(f)ij} = \begin{bmatrix} \mathbf{x}'_{(f)ij1} \\ \mathbf{x}'_{(f)ij2} \\ \vdots \\ \mathbf{x}'_{(f)ijn_{ij}} \end{bmatrix}, \quad \mathbf{X}_{(3)ij} = \begin{bmatrix} \mathbf{x}'_{(3)ij1} \\ \mathbf{x}'_{(3)ij2} \\ \vdots \\ \mathbf{x}'_{(3)ijn_{ij}} \end{bmatrix}, \quad \text{and}$$

$$\mathbf{X}_{(2)ij} = \begin{bmatrix} \mathbf{x}'_{(2)ij1} \\ \mathbf{x}'_{(2)ij2} \\ \vdots \\ \mathbf{x}'_{(2)ijn_{ij}} \end{bmatrix}.$$

Under the distributional assumptions given above, it follows that

$$\mathsf{E}(\mathbf{y}_i) = \mathbf{X}_{(f)i}\boldsymbol{\beta}, \quad (2.14)$$

where

$$\mathbf{X}_{(f)i} = \begin{bmatrix} \mathbf{X}_{(f)i1} \\ \mathbf{X}_{(f)i2} \\ \vdots \\ \mathbf{X}_{(f)in_i} \end{bmatrix}.$$

Also

$$\mathsf{Cov}(\mathbf{y}_i, \mathbf{y}'_i) = \mathbf{X}_{(3)i}\boldsymbol{\Phi}_{(3)}\mathbf{X}'_{(3)i} + \boldsymbol{\Lambda}_i, \quad (2.15)$$

where

$$\mathbf{\Lambda}_i = \begin{bmatrix} \mathbf{\Lambda}_{i1} + \mathbf{D}_{i1} & 0 & \ldots & 0 \\ 0 & \mathbf{\Lambda}_{i2} + \mathbf{D}_{i2} & \ldots & 0 \\ \vdots & \vdots & \ddots & \vdots \\ 0 & 0 & \ldots & \mathbf{\Lambda}_{in_i} + \mathbf{D}_{in_i} \end{bmatrix} \quad (2.16)$$

with

$$\mathbf{\Lambda}_{ij} = \mathbf{X}_{(2)ij} \mathbf{\Phi}_{(3)} \mathbf{X}'_{(2)ij}$$

and

$$\mathbf{D}_{ij} = \begin{bmatrix} \lambda_{ij1} & 0 & \ldots & 0 \\ 0 & \lambda_{ij2} & \ldots & 0 \\ \vdots & \vdots & \ddots & \vdots \\ 0 & 0 & \ldots & \lambda_{ijn_{ij}} \end{bmatrix} \quad (2.17)$$

with

$$\lambda_{ijk} = \mathbf{x}'_{(1)ijk} \mathbf{\Phi}_{(1)} \mathbf{x}_{(1)ijk} \ .$$

2.2.4 Parameter estimation

The model given by (2.13) introduced in the Section *A general level-3 model* (page 19) may be written as

$$\mathbf{y}_i = \mathbf{X}_{(f)i}\boldsymbol{\beta} + \mathbf{X}_{(3)i}\mathbf{v}_i + \sum_{j=1}^{n_i} \mathbf{Z}_{(2)ij}\mathbf{u}_{ij} + \sum_{j=1}^{n_i}\sum_{k=1}^{n_{ij}} \mathbf{U}_{(1)ijk}\mathbf{e}_{ijk} \quad (2.18)$$

where

2.2 THEORETICAL BACKGROUND

$$\mathbf{X}_{(3)i} = \begin{bmatrix} \mathbf{X}_{(3)i1} \\ \mathbf{X}_{(3)i2} \\ \vdots \\ \mathbf{X}_{(3)in_i} \end{bmatrix}, \qquad (2.19)$$

$$\mathbf{Z}_{(2)ij} = \begin{bmatrix} \mathbf{0} \\ \vdots \\ \mathbf{0} \\ \mathbf{X}_{(2)ij} \\ \mathbf{0} \\ \vdots \\ \mathbf{0} \end{bmatrix}, \qquad (2.20)$$

$$\mathbf{U}_{(1)ijk} = \begin{bmatrix} \mathbf{0} \\ \vdots \\ \mathbf{0} \\ \mathbf{X}_{(1)ijk} \\ \mathbf{0} \\ \vdots \\ \mathbf{0} \end{bmatrix}, \qquad (2.21)$$

and where $\mathbf{v}_i$, $\mathbf{u}_{ij}$, and $\mathbf{e}_{ijk}$ denote the random parameter vectors on level 3, level 2, and level 1 of the model. It will be convenient to replace the double subscript jk with the single subscript l where $l = 1, 2, \ldots, n_i^*$ with

$$n_i^* = \sum_{j=1}^{n_i} n_{ij} .$$

Thus, (2.18) can be rewritten as

$$\mathbf{y}_i = \mathbf{X}_{(3)i}\mathbf{v}_i + \sum_{j=1}^{n_i} \mathbf{Z}_{(2)ij}\mathbf{u}_{ij} + \sum_{l=1}^{n_i^*} \mathbf{U}_{(1)il}\mathbf{e}_{il} . \qquad (2.22)$$

Under the distributional assumptions given in Section *A general level-3 model*, it follows that

$$E(\mathbf{y}_i) = \mathbf{X}_{(f)i}\boldsymbol{\beta},$$

where

$$\mathbf{X}_{(f)i} = \begin{bmatrix} \mathbf{X}_{(f)i1} \\ \mathbf{X}_{(f)i2} \\ \vdots \\ \mathbf{X}_{(f)in_i} \end{bmatrix}. \qquad (2.23)$$

It also follows that

$$\begin{aligned}\text{Cov}(\mathbf{y}_i, \mathbf{y}'_i) &= \boldsymbol{\Sigma}_i \\ &= \mathbf{X}_{(3)i}\boldsymbol{\Phi}_{(3)}\mathbf{X}'_{(3)i} + \sum_{j=1}^{n_i} \mathbf{Z}_{(2)ij}\boldsymbol{\Phi}_{(2)}\mathbf{Z}'_{(2)ij} + \\ &\quad + \sum_{l=1}^{n_i^*} \mathbf{U}_{(1)il}\boldsymbol{\Phi}_{(1)}\mathbf{U}'_{(1)il}.\end{aligned}$$

Suppose that $\hat{\boldsymbol{\Phi}}_{(3)}$, $\hat{\boldsymbol{\Phi}}_{(2)}$, and $\hat{\boldsymbol{\Phi}}_{(1)}$ are consistent estimators of $\boldsymbol{\Phi}_{(3)}$, $\boldsymbol{\Phi}_{(2)}$, and $\boldsymbol{\Phi}_{(1)}$, respectively, so that

$$\mathbf{V}_i = \mathbf{X}_{(3)i}\hat{\boldsymbol{\Phi}}_{(3)}\mathbf{X}'_{(3)i} + \sum_{j=1}^{n_i} \mathbf{Z}_{(2)ij}\hat{\boldsymbol{\Phi}}_{(2)}\mathbf{Z}'_{(2)ij} + \sum_{l=1}^{n_i^*} \mathbf{U}_{(1)il}\hat{\boldsymbol{\Phi}}_{(1)}\mathbf{U}'_{(1)il}$$

is a consistent estimator of $\boldsymbol{\Sigma}_i$.

The generalized least squares estimator $\hat{\boldsymbol{\beta}}$ of $\boldsymbol{\beta}$ is obtained as the minimum of the quadratic function

2.2 THEORETICAL BACKGROUND

$$\mathbf{Q}_f = \sum_{i=1}^{N}[\mathbf{y}_i - \mathbf{X}_{(f)i}\boldsymbol{\beta}]'\mathbf{V}_i^{-1}[\mathbf{y}_i - \mathbf{X}_{(f)i}\boldsymbol{\beta}]$$

with solution

$$\hat{\boldsymbol{\beta}} = \left[\sum_{i=1}^{N}\mathbf{X}'_{(f)i}\mathbf{V}_i^{-1}\mathbf{X}_{(f)i}\right]^{-1}\left[\sum_{i=1}^{N}\mathbf{X}'_{(f)i}\mathbf{V}_i^{-1}\mathbf{y}_i\right]. \qquad (2.24)$$

In order to estimate $\boldsymbol{\Phi}_{(3)}$, $\boldsymbol{\Phi}_{(2)}$, and $\boldsymbol{\Phi}_{(1)}$ let

$$\mathbf{y}_i^* = \mathsf{vecs}\,(\mathbf{y}_i - \mathbf{X}_{(f)i}\boldsymbol{\beta})(\mathbf{y}_i - \mathbf{X}_{(f)i}\boldsymbol{\beta})', \qquad (2.25)$$

then

$$\mathsf{E}(\mathbf{y}_i^*) = \mathsf{vecs}\,\boldsymbol{\Sigma}_i\,.$$

Using the result (Browne, 1974), on vector operations, *viz*,

$$\mathsf{vec}\,(\mathbf{CAC}') = (\mathbf{C}\otimes\mathbf{C})\mathsf{vec}\mathbf{A}\,,$$

it follows that

$$\begin{aligned}
\mathsf{vec}\mathbf{V}_i &= \mathbf{X}_{(3)i}\otimes\mathbf{X}_{(3)i}\mathsf{vec}\,\boldsymbol{\Phi}_{(3)} + \sum_{j=1}^{n_i}(\mathbf{Z}_{(2)ij}\otimes\mathbf{Z}_{(2)ij})\mathsf{vec}\,\boldsymbol{\Phi}_{(2)} + \\
&\quad + \sum_{l=1}^{n_i^*}(\mathbf{U}_{(1)il}\otimes\mathbf{U}_{(1)il})\mathsf{vec}\,\boldsymbol{\Phi}_{(1)}\,.
\end{aligned}$$

There exists a unique matrix (see Browne, 1974, or McCulloch, 1982) $\mathbf{G}_p : p^2 \times \frac{1}{2}p(p+1)$ such that

$$\mathsf{vec}\,\mathbf{A} = \mathbf{G}_p\mathsf{vecs}\,\mathbf{A}$$

with $\mathbf{A}$ a symmetric $p \times p$ matrix. There is also a non-unique matrix $\mathbf{H}_p : \frac{1}{2}p(p+1) \times p^2$ such that

$$\text{vecs } \mathbf{A} = \mathbf{H}_p \text{vec } \mathbf{A}$$

The vector vecs $\boldsymbol{\Sigma}_i$, consisting of the non-duplicated elements of $\boldsymbol{\Sigma}_i$, can then be written as

$$\text{vecs } \boldsymbol{\Sigma}_i = \mathbf{X}_i^* \boldsymbol{\tau}$$

where

$$\mathbf{X}_i^{*\prime} = \mathbf{H}_{n_i^*} \begin{bmatrix} (\mathbf{X}_{(3)i} \otimes \mathbf{X}_{(3)i})\mathbf{G}_q \\ \left(\sum_{j=1}^{n_i} \mathbf{Z}_{(2)ij} \otimes \mathbf{Z}_{(2)ij}\right)\mathbf{G}_m \\ \left(\sum_{l=1}^{n_i^*} \mathbf{U}_{(1)il} \otimes \mathbf{U}_{(1)il}\right)\mathbf{G}_r \end{bmatrix} \quad (2.26)$$

and

$$\boldsymbol{\tau} = \begin{bmatrix} \text{vecs } \boldsymbol{\Phi}_{(3)} \\ \text{vecs } \boldsymbol{\Phi}_{(2)} \\ \text{vecs } \boldsymbol{\Phi}_{(1)} \end{bmatrix}. \quad (2.27)$$

Now consider the quadratic form

$$\mathbf{Q}_T = \sum_{i=1}^{N} \left\{ [\mathbf{y}_i^* - \mathbf{X}_i^* \boldsymbol{\tau}]' \mathbf{W}_i^{-1} [\mathbf{y}_i^* - \mathbf{X}_i^* \boldsymbol{\tau}] \right\}$$

where $\mathbf{W}_i$ is a consistent estimator of the covariance of $\mathbf{y}_i^*$.

It has, for example, been shown by Browne (1974) and Goldstein (1989) that, if

$$\mathbf{W}_i^{-1} = \frac{1}{2} \mathbf{G}_{n_i}' \left(\mathbf{V}_i^{-1} \otimes \mathbf{V}_i^{-1} \right) \mathbf{G}_{n_i}, \quad (2.28)$$

2.2 THEORETICAL BACKGROUND

then $\mathbf{W}_i$ is a consistent estimator of the covariance of $\mathbf{y}_i^*$.

Minimization of $\mathbf{Q}_T$ with respect to τ yields

$$\hat{\tau} = \left[\sum_{i=1}^{N} \mathbf{X}_i^{*\prime} \mathbf{W}_i^{-1} \mathbf{X}_i^*\right]^{-1} \left[\sum_{i=1}^{N} \mathbf{X}_i^{*\prime} \mathbf{W}_i^{-1} \mathbf{y}_i^*\right] \qquad (2.29)$$

In order to ensure computational efficiency, the components of $\hat{\tau}$ are further simplified as discussed in du Toit (1995).

IGLS estimators

The IGLS estimators $\hat{\beta}$ of β and $\hat{\tau}$ of τ can be obtained as follows:

(i) Set $\mathbf{V} = \mathbf{I}$
(ii) Calculate β_k (cf. (2.24))
(iii) Calculate $\mathbf{y}_i^* = \text{vecs}\,(\mathbf{y}_i - \mathbf{X}_{(3)i}\beta_k)(\mathbf{y}_i - \mathbf{X}_{(3)i}\beta_k)'$ (cf. (2.25))
(iv) Calculate τ_k by initially setting $\Phi_{(1)} = \mathbf{I}$, $\Phi_{(2)} = \mathbf{I}$, and $\Phi_{(3)} = \mathbf{I}$
(v) Update $\mathbf{V}_i$ (cf. (2.28))

Repeat steps (ii) to (v) until convergence is obtained. The algorithm described above is known as Iterative Generalized Least Squares (IGLS).

2.2.5 Statistical inference

In the following section (*Standard errors*), results are given which are required for the calculation of the standard errors of the estimated parameters.

Next, in *Contrasts*, we discuss hypotheses of the form $c_1\beta_1 + c_2\beta_2 + \cdots + c_q\beta_q = k$ about the elements of the fixed parameter vector β.

We continue with the calculation of empirical Bayes estimates (p. 30) and conclude with discussing likelihood ratio tests (p. 31).

Standard errors

From (2.24) it follows that the covariance matrix of $\hat{\beta}$ is given by

$$\text{Cov}(\hat{\beta}, \hat{\beta}') = \left[\mathbf{X}_{(f)}^{*\prime} \mathbf{\Sigma}^{-1} \mathbf{X}_{(f)}^{*} \right]^{-1}. \qquad (2.30)$$

In practice, $\mathbf{\Sigma}_i$ is unknown and is replaced by the maximum likelihood estimator $\mathbf{\Sigma}_i(\hat{\tau}) = \mathbf{V}_i$. Hence, a consistent estimate of the covariance matrix of $\hat{\beta}$ is given by

$$\text{Cov}(\hat{\beta}, \hat{\beta}') = \left[\mathbf{X}_{(f)}^{*\prime} \mathbf{V}^{-1} \mathbf{X}_{(f)}^{*} \right]^{-1}. \qquad (2.31)$$

Similarly, it can be shown that a consistent estimator of the covariance matrix of $\hat{\tau}$ (cf. (2.29)) is given by

$$\text{Cov}(\hat{\tau}, \hat{\tau}') = \left[\sum_{i=1}^{N} \mathbf{X}_i^{*\prime} \mathbf{W}_i^{-1} \mathbf{X}_i^{*} \right]^{-1}. \qquad (2.32)$$

The diagonal elements of the covariance matrix (2.31) may be used to obtain large-sample estimates of the standard errors for the fixed parameter estimates. For large samples, $\hat{\beta}$ and $\hat{\tau}$ have approximate multivariate normal distributions. See, for example, Malinvaud (1970) for general results on the distribution of least squares estimators.

Contrasts

The construction of contrasts or linear functions of the parameters is a useful statistical analysis tool and enables the researcher to perform hypothesis testing concerning the equality of subsets of parameters. In this section a summary of the results required for contrast testing is given.

A $p \times q$ contrast matrix $\mathbf{C}$, where p denotes the number of contrasts, can be used to formulate a complex hypothesis about several elements of β. The hypothesis is written in the form $\mathbf{C}\beta = \mathbf{k}$, where $\mathbf{k}$ is a known $p \times 1$ vector.

2.2 THEORETICAL BACKGROUND

Consider as an example the case where $q = 3$ and the following hypothesis is to be tested:

$$\beta_1 - \beta_2 = 0$$
$$\beta_3 - \beta_2 = 0 .$$

The null hypothesis can be formulated as

$$\begin{bmatrix} 1 & -1 & 0 \\ 0 & -1 & 1 \end{bmatrix} \begin{bmatrix} \beta_1 \\ \beta_2 \\ \beta_3 \end{bmatrix} = \begin{bmatrix} 0 \\ 0 \end{bmatrix} .$$

For large samples, it can be shown that the vector variate $\mathbf{C}\hat{\boldsymbol{\beta}}$ will be approximately distributed as $N(\mathbf{C}\boldsymbol{\beta}, \mathbf{C}(\mathbf{X}'_{(f)}\hat{\mathbf{V}}^{-1}\mathbf{X}_{(f)})^{-1}\mathbf{C}')$.

Therefore, if H_0 is true,

$$\mathbf{M} = (\mathbf{C}\hat{\boldsymbol{\beta}} - \mathbf{k})' \left\{ \mathbf{C}(\mathbf{X}'_{(f)}\hat{\mathbf{V}}^{-1}\mathbf{X}_{(f)})^{-1}\mathbf{C}' \right\} (\mathbf{C}\hat{\boldsymbol{\beta}} - \mathbf{k}) \qquad (2.33)$$

has an approximate χ^2 distribution with p degrees of freedom.

Let $\mathbf{c}'$ denote the i-th row of $\mathbf{C}$ and $\chi^2_{q,\alpha}$ the critical value of the χ^2 distribution with q degrees of freedom. A set of $100(1-\alpha)\%$ simultaneous confidence intervals for the p elements of $\mathbf{C}\boldsymbol{\beta}$ is given by the p intervals

$$\mathbf{c}'_i\hat{\boldsymbol{\beta}} \pm \left\{ \mathbf{c}'_i(\mathbf{X}'_{(j)}\hat{\mathbf{V}}^{-1}\mathbf{X}_{(j)})^{-1}\mathbf{c}_i \chi^2_{q,\alpha} \right\}^{0.5}, \quad p < q . \qquad (2.34)$$

The null hypothesis $H_0 : \hat{\beta}_j = 0$, $j = 1, 2, \ldots, q$ is tested by using the test statistic

$$z = \frac{\hat{\beta}_k}{\text{S.E.}(\hat{\beta}_k)} .$$

which, for large samples, has an approximate $N(0,1)$ distribution if H_0 is true.

Empirical Bayes estimates

The residuals $\hat{\mathbf{v}}_i$, $\hat{\mathbf{u}}_i$, and $\hat{\mathbf{e}}_i$ (see Section 2.2.3) may be estimated as follows.

Let

$$\tilde{\mathbf{y}}_i = \mathbf{y}_i - \mathbf{X}_{(f)i}\boldsymbol{\beta} ,$$

Under the assumption of multivariate normality it follows that $\tilde{\mathbf{y}}_i$ is approximately distributed as $N(\mathbf{0}, \boldsymbol{\Sigma}_i)$ and $\mathbf{v}_i$ is approximately distributed as $N(\mathbf{0}, \boldsymbol{\Phi}_{(3)})$, and hence the joint distribution of $\tilde{\mathbf{y}}_i$ and $\mathbf{v}_i$ is

$$\begin{pmatrix} \tilde{\mathbf{y}}_i \\ \mathbf{v}_i \end{pmatrix} \sim N \left(\begin{pmatrix} \mathbf{0} \\ \mathbf{0} \end{pmatrix} , \begin{matrix} \boldsymbol{\Sigma} & \mathbf{X}_{(3)i}\boldsymbol{\Phi}_{(3)} \\ \boldsymbol{\Phi}_{(3)}\mathbf{X}'_{(3)i} & \boldsymbol{\Phi}_{(3)} \end{matrix} \right).$$

From standard results on conditional distributions (see for example Morrison, 1991) it follows that

$$\begin{aligned} \mathsf{E}(\mathbf{v}_i|\tilde{\mathbf{y}}_i) &= \mathbf{0} + (\mathbf{X}_{(3)i}\boldsymbol{\Phi}_{(3)})'\boldsymbol{\Sigma}_i^{-1}(\tilde{\mathbf{y}}_i - \mathbf{0}) \\ &= (\mathbf{X}_{(3)i}\boldsymbol{\Phi}_{(3)})'\boldsymbol{\Sigma}_i^{-1}\tilde{\mathbf{y}}_i \\ &= \boldsymbol{\Phi}_{(3)}\mathbf{X}'_{(3)i}\boldsymbol{\Sigma}_i^{-1}\tilde{\mathbf{y}}_i , \end{aligned}$$

Thus, the empirical Bayes estimate of $\mathbf{v}_i$ is

$$\hat{\mathbf{v}}_i = \hat{\boldsymbol{\Phi}}_{(3)}\mathbf{X}'_{(3)i}\mathbf{V}_i^{-1}(\mathbf{y}_i - \mathbf{X}_{(f)i}\hat{\boldsymbol{\beta}}) . \tag{2.35}$$

Similarly,

$$\hat{\mathbf{u}}_{ij} = \hat{\boldsymbol{\Phi}}_{(2)}\mathbf{X}'_{(2)ij}\mathbf{V}_i^{-1}(\mathbf{y}_{ij} - \mathbf{X}_{(f)ij}\hat{\boldsymbol{\beta}}) . \tag{2.36}$$

Likelihood ratio tests

Finally, likelihood ratio tests are considered. Tests of a null hypothesis against a restricted alternative hypothesis can be constructed, provided that two conditions are met. Firstly, the models under H_0 and H_1 should be estimable and secondly, the parameter space Ω_0 for H_0 must be a subset of the parameter space Ω of H_1.

Use is made of the likelihood ratio test statistic

$$\lambda = \frac{L_0(\hat{\beta}, \hat{\tau})}{L_1(\hat{\beta}, \hat{\tau})}, \qquad (2.37)$$

where L_0 and L_1 denote the likelihood functions under H_0 and H_1, respectively. For N large (see for example Anderson, 1984), $-2\ln\lambda$ has an approximate $\chi^2(v)$ distribution where the number of degrees of freedom v is the difference in the number of parameters estimated under H_1 and the number of parameters estimated under H_0.

Example:

Consider the null hypothesis

$$H_0: \text{Cov}(\mathbf{y}_{ij}, \mathbf{y}'_{ij}) = \mathbf{X}_{(2)ij} \boldsymbol{\Phi}_{(2)} \mathbf{X}'_{(2)ij} + \boldsymbol{\Phi}_{(1)} \mathbf{I}_{n_{ij}}$$

as opposed to the alternative hypothesis

$$H_1: \text{Cov}(\mathbf{y}_{ij}, \mathbf{y}'_{ij}) = \mathbf{X}_{(3)ij} \boldsymbol{\Phi}_{(3)} \mathbf{X}'_{(3)ij} + \mathbf{X}_{(2)ij} \boldsymbol{\Phi}_{(2)} \mathbf{X}'_{(2)ij} + \boldsymbol{\Phi}_{(1)} \mathbf{I}_{n_{ij}}.$$

Let

$$\lambda = \frac{L_0(\hat{\boldsymbol{\Phi}}_{(1)}, \hat{\boldsymbol{\Phi}}_{(2)})}{L_1(\hat{\boldsymbol{\Phi}}_{(1)}, \hat{\boldsymbol{\Phi}}_{(2)}, \hat{\boldsymbol{\Phi}}_{(3)})}.$$

For N large, $-2\ln\lambda = -2(\ln L_0 - \ln L_1)$ has an approximate $\chi^2(v)$ distribution with the number of degrees of freedom, $v = \frac{1}{2}q(q+1)$, which is the number of non-duplicated elements of $\Phi_{(3)}$. Note that $\ln L$ is the log-likelihood function

$$\ln L = -\frac{1}{2}\sum_{i=1}^{N}\left\{n_i\ln(2\pi) + \ln|\Sigma_i| + \text{tr}\Sigma_i^{-1}(\mathbf{y}_i - \mathbf{X}_{(f)i}\boldsymbol{\beta})(\mathbf{y}_i - \mathbf{X}_{(f)i}\boldsymbol{\beta})'\right\}$$

with $\mathbf{X}_i^*\boldsymbol{\beta}$ and Σ_i, respectively, the expected value and covariance of $\mathbf{y}_i$.

2.2.6 Multilevel logit models

Introduction

Survey data usually consist of a mixture of biographical, geographical, and response variables and frequently have a hierarchical structure. Quite often, these response variables are categorical in nature. In the last few decades a wide variety of methods for the analysis of categorical data have been proposed. Many of these are generalizations of continuous data analysis methods (see for example Bishop, Fienberg, & Holland, 1975, and Agresti, 1990). In order to accommodate the structure of hierarchical data, a multilevel modeling approach may be used.

A level-3 logit model

To introduce a logit model with a categorical response variable, consider the following frequency table, where the subscripts ijk refer to subpopulation k within the i-th level-3 and j-th level-2 combination.

For illustrative purposes, it is assumed that a maximum of six subpopulations are formed according to the gender and age of respondents, as shown below for the i-th level-3 and j-th level-2 unit.

2.2 THEORETICAL BACKGROUND

Gender	Age	Number of responses			Total
		Negative	Don't know	Yes	
Male	18 to 29 years	$f_{ij1,1}$	$f_{ij1,2}$	$f_{ij1,3}$	$f_{ij1\cdot}$
Male	30 to 49 years	$f_{ij2,1}$	$f_{ij2,2}$	$f_{ij2,3}$	$f_{ij2\cdot}$
Male	50 years and older	$f_{ij3,1}$	$f_{ij3,2}$	$f_{ij3,3}$	$f_{ij3\cdot}$
Female	18 to 29 years	$f_{ij4,1}$	$f_{ij4,2}$	$f_{ij4,3}$	$f_{ij4\cdot}$
Female	30 to 49 years	$f_{ij5,1}$	$f_{ij5,2}$	$f_{ij5,3}$	$f_{ij5\cdot}$
Female	50 years and older	$f_{ij6,1}$	$f_{ij6,2}$	$f_{ij6,3}$	$f_{ij6\cdot}$

The frequencies $f_{ijk,1}$, $f_{ijk,2}$, and $f_{ijk,3}$ denote the number of negative, don't know, and positive responses, respectively, while $f_{ijk\cdot}$ is the total number of responses for the k-th subpopulation, $k = 1, 2, \ldots, 6$. The number of response categories is denoted by c which, in this case, is equal to three.

The vector of responses is formed by using the third category, i.e., the number of positive responses, as reference category. For each of the six subpopulations described above, two elements of the vector of responses are formed as follows:

$$\mathbf{y}_{ij} = \left[\ln\frac{f_{ij1,1}}{f_{ij1,3}} \quad \ln\frac{f_{ij1,2}}{f_{ij1,3}} \quad \cdots \quad \ln\frac{f_{ij6,1}}{f_{ij6,3}} \quad \ln\frac{f_{ij6,2}}{f_{ij6,3}} \right]. \tag{2.38}$$

For each subpopulation k within the i-th level-3 and j-th level-2 unit, the probability π_{ijkl} of the l-th response ($l = 1, 2$) occurring is estimated by $p_{ijkl} = \frac{f_{ijk,l}}{f_{ijk\cdot}}$. Hence, a typical element of $\mathbf{y}_{ij}$ is $\ln\frac{p_{ijkl}}{p_{ijkc}}$.

From this it follows that

$$\mathbf{y}_i = f(\mathbf{p}_{ij}) \tag{2.39}$$

where

$$\mathbf{p}'_{ij} = \left[p_{ij1,1} \; p_{ij1,2} \quad \cdots \quad p_{ij6,1} \; p_{ij6,2} \right].$$

It is assumed that the expected value of the response vector $\mathbf{y}_{ij}$ can be expressed in the form

$$E(\mathbf{y}_{ij}) = \mathbf{X}_{(f)ij}\boldsymbol{\beta} \qquad (2.40)$$

where the elements of $\mathbf{X}_{(f)ij}$ depend on whether provision is made for the inclusion of an intercept or constant effect and on the way in which subpopulations are formed. The elements of the vector $\boldsymbol{\beta}$ are fixed, but unknown, parameters to be estimated.

Denote the covariance matrix of $\mathbf{y}_{ij}$ by $\boldsymbol{\Sigma}_{ij}$. An approximate expression for $\boldsymbol{\Sigma}_i$ is obtained through use of the following first order Taylor expansion of $\mathbf{y}_{ij}$ evaluated in the neighborhood of $\boldsymbol{\pi}_{ij}$ (see for example Cramer, 1946)

$$\mathbf{y}_{ij} \simeq f(\boldsymbol{\pi}_{ij}) + \mathbf{J}_{ij}(\mathbf{p}_{ij} - \boldsymbol{\pi}_{ij}) \qquad (2.41)$$

where

$$\mathbf{J}_{ij} : s \times cs = \frac{\partial \mathbf{y}_{ij}}{\partial \mathbf{y}'_{ij}}|\mathbf{p}_{ij} = \boldsymbol{\pi}_{ij},$$

c is the number of categories of the response variable and s denotes the total number of subpopulations in the (i,j)-th unit.

The covariance of $\mathbf{y}_{ij}$, to the first order of approximation, can be written as

$$\boldsymbol{\Sigma}_{ij} = \mathbf{J}_{ij}\,\text{Cov}\,(\mathbf{p}_{ij}, \mathbf{p}'_{ij})\mathbf{J}'_{ij} = \mathbf{J}_{ij}\mathbf{V}_{ij}\mathbf{J}'_{ij}. \qquad (2.42)$$

It can be shown that, if the c-th category of the response variable is used as the reference category, $\boldsymbol{\Sigma}_{ij}$ has, to the first order of approximation, the following form

$$\boldsymbol{\Sigma}_{ij} = \begin{bmatrix} \frac{1}{f_{ij1.}}\boldsymbol{\Psi}_{11} & 0 & \cdots & 0 \\ 0 & \frac{1}{f_{ij2.}}\boldsymbol{\Psi}_{22} & \cdots & 0 \\ \vdots & \vdots & \ddots & \vdots \\ 0 & 0 & \cdots & \frac{1}{f_{ijs.}}\boldsymbol{\Psi}_{ss} \end{bmatrix} \qquad (2.43)$$

2.2 THEORETICAL BACKGROUND

For a c category response variable

$$\Psi_{kk} = \mathbf{D}\pi_{ijk} + \frac{\mathbf{jj}'}{\pi_{ijk,c}}, \qquad (2.44)$$

where $\pi_{ijk,c}$ is the probability that respondents in the k-th subpopulation from the i-th level-3 and j-th level-2 unit will select the c-th category of the response variable.

In order to accommodate the level-1 error structure, the model can be written as

$$\mathbf{y}_{ij} = \mathbf{X}_{(f)ij}\boldsymbol{\beta} + \mathbf{X}_{(1)ij}\mathbf{e}_{ij}. \qquad (2.45)$$

It is assumed that $\mathbf{e}_{i1}, \mathbf{e}_{i2}, \ldots, \mathbf{e}_{in_i}$ are $i.i.d.$ with mean $\mathbf{0}$ and covariance matrix $\boldsymbol{\Phi}_{(1)}$. Assuming that the number of response categories is c, the matrix $\mathbf{X}_{(1)ij}$ may be regarded as an appropriate weight matrix so that $\mathbf{X}_{(1)ij}\boldsymbol{\Phi}_{(1)}\mathbf{X}'_{(1)ij}$ has the form (2.43), where

$$\mathbf{X}_{(1)ij} = \mathbf{D}_{ij} = \begin{bmatrix} \frac{1}{\sqrt{f_{ij1\cdot}}}\mathbf{I}_{c-1} & 0 & 0 & 0 & 0 & 0 \\ 0 & \frac{1}{\sqrt{f_{ij2\cdot}}}\mathbf{I}_{c-1} & 0 & 0 & 0 & 0 \\ 0 & 0 & \frac{1}{\sqrt{f_{ij3\cdot}}}\mathbf{I}_{c-1} & 0 & 0 & 0 \\ 0 & 0 & 0 & \frac{1}{\sqrt{f_{ij4\cdot}}}\mathbf{I}_{c-1} & 0 & 0 \\ 0 & 0 & 0 & 0 & \frac{1}{\sqrt{f_{ij5\cdot}}}\mathbf{I}_{c-1} & 0 \\ 0 & 0 & 0 & 0 & 0 & \frac{1}{\sqrt{f_{ij6\cdot}}}\mathbf{I}_{c-1} \end{bmatrix}, \qquad (2.46)$$

and the following constraints are imposed on the level-1 covariance matrix:

$$\boldsymbol{\Phi}_{(1)} = \begin{bmatrix} \boldsymbol{\Phi}_{(1)11} & 0 & 0 & 0 & 0 & 0 \\ 0 & \boldsymbol{\Phi}_{(1)22} & 0 & 0 & 0 & 0 \\ 0 & 0 & \boldsymbol{\Phi}_{(1)33} & 0 & 0 & 0 \\ 0 & 0 & 0 & \boldsymbol{\Phi}_{(1)44} & 0 & 0 \\ 0 & 0 & 0 & 0 & \boldsymbol{\Phi}_{(1)55} & 0 \\ 0 & 0 & 0 & 0 & 0 & \boldsymbol{\Phi}_{(1)66} \end{bmatrix}. \qquad (2.47)$$

The model given in (2.45) is further extended to include random components on level 2 and level 3 of the model as follows

$$\begin{aligned} \mathbf{y}_{ij} &= \mathbf{X}_{(f)ij}\left[\boldsymbol{\beta} + \mathbf{S}_{(3)ij}\mathbf{v}_i + \mathbf{S}_{(2)ij}\mathbf{u}_{ij}\right] + \mathbf{X}_{(1)ij}\mathbf{e}_{ij} \qquad (2.48) \\ &= \mathbf{X}_{(f)ij}\boldsymbol{\beta} + \mathbf{X}_{(3)ij}\mathbf{v}_i + \mathbf{X}_{(2)ij}\mathbf{u}_{ij} + \mathbf{X}_{(1)ij}\mathbf{e}_{ij} \,. \end{aligned}$$

The vector $\mathbf{v}_i$ represents the coefficients of variables allowed to be random on level 3 and it is assumed that $\mathbf{v}_1, \mathbf{v}_2, \ldots, \mathbf{v}_N$ are independently and identically distributed with mean $\mathbf{0}$ and covariance matrix $\boldsymbol{\Phi}_{(3)}$. It is assumed that $\mathbf{v}_i$, $\mathbf{u}_{ij}$, and $\mathbf{e}_{ijk}$ are independent.

$\mathbf{S}_{(2)}$ is a $t \times m$ ($m \leq t$) matrix formed by the selection of columns from the identity matrix of order t. These columns correspond to those elements of $\mathbf{b}_{ij}$ that are allowed to be random on level 2. If, for example, $t = 4$ and only the first and fourth coefficients are allowed to vary randomly on level 2, then

$$\mathbf{S}_{(2)} = \begin{bmatrix} 1 & 0 \\ 0 & 0 \\ 0 & 0 \\ 0 & 1 \end{bmatrix}.$$

The matrix $\mathbf{S}_{(3)}$ is defined in a similar way.

Let

$$\mathbf{y}'_i = \begin{bmatrix} \mathbf{y}'_{i1} & \cdots & \mathbf{y}'_{ij} & \cdots & \mathbf{y}'_{in_i} \end{bmatrix}$$

and denote the covariance matrix of $\mathbf{y}_i$ by $\boldsymbol{\Sigma}_i$. It follows that

$$\boldsymbol{\Sigma}_i = \mathbf{X}_{(3)ij}\boldsymbol{\Phi}_{(3)}\mathbf{X}'_{(3)ij} + \boldsymbol{\Lambda}_i \,, \qquad (2.49)$$

where

2.2 THEORETICAL BACKGROUND

$$\Lambda_i = \begin{bmatrix} \Lambda_{i1} & 0 & \cdots & 0 \\ 0 & \Lambda_{i2} & \cdots & 0 \\ \vdots & \vdots & \ddots & \vdots \\ 0 & 0 & \cdots & \Lambda_{in_j} \end{bmatrix}$$

with

$$\Lambda_{ij} = \mathbf{X}_{(2)ij}\Phi_{(2)}\mathbf{X}'_{(2)ij} + \mathbf{X}_{(1)ij}\Phi_{(1)}\mathbf{X}'_{(1)ij}.$$

Estimation considerations

In this section it is shown that parameter estimation of a level-3 logit model is achieved by fitting a general level-3 model with random coefficients on level 3 and level 2 of the hierarchy.

Let

$$\mathbf{y}_{ijk} = \begin{bmatrix} y_{ijk,1} \\ y_{ijk,2} \\ \vdots \\ y_{ijk,c-1} \end{bmatrix} \qquad (2.50)$$

denote a set of $c-1$ responses from subpopulation k, $k = 1, 2, \ldots, s$, from the j-th level-2 and i-th level-3 unit.

It follows from (2.48) that

$$\begin{aligned}
y_{ijk,1} &= \mathbf{x}'_{(f)ijk,1}\boldsymbol{\beta} + \mathbf{x}'_{(3)ijk,1}\mathbf{v}_i + \mathbf{x}'_{(2)ijk,1}\mathbf{u}_{ij} + \mathbf{x}'_{(1)ijk,1}\mathbf{e}_{ijk} \\
y_{ijk,2} &= \mathbf{x}'_{(f)ijk,2}\boldsymbol{\beta} + \mathbf{x}'_{(3)ijk,2}\mathbf{v}_i + \mathbf{x}'_{(2)ijk,2}\mathbf{u}_{ij} + \mathbf{x}'_{(1)ijk,2}\mathbf{e}_{ijk} \\
\vdots \quad &\quad \vdots \\
y_{ijk,c-1} &= \mathbf{x}'_{(f)ijk,c-1}\boldsymbol{\beta} + \mathbf{x}'_{(3)ijk,c-1}\mathbf{v}_i + \mathbf{x}'_{(2)ijk,c-1}\mathbf{u}_{ij} + \mathbf{x}'_{(1)ijk,c-1}\mathbf{e}_{ijk}
\end{aligned} \qquad (2.51)$$

Under the assumption of a multinomial level-1 error structure (see the previous Section), it follows that the elements of $\mathbf{e}_{ijk}$ are correlated.

A multilevel analysis of logit models requires that

(i) $\mathbf{e}_{ijk}$, $k = 1, 2, \ldots, s_j$ are independently distributed and that
(ii) $\text{Cov}(\mathbf{e}_{ijk}, \mathbf{e}'_{ijk}) = \mathbf{\Phi}_{(1)kk}$ (cf. (2.43) and (2.46))

which implies that the $c - 1$ elements of $\mathbf{e}_{ijk}$ are correlated, but that $\mathbf{e}_{ijk}$ and $\mathbf{e}_{ijm}$, $k \neq m$, are uncorrelated.

Let $\mathbf{X}_{(f)ijk}$, $\mathbf{X}_{(3)ijk}$, $\mathbf{X}_{(2)ijk}$, and $\mathbf{X}_{(1)ijk}$ have typical rows

$$\mathbf{x}'_{(f)ijk,l}, \ \mathbf{x}'_{(3)ijk,l}, \ \mathbf{x}'_{(2)ijk,l}, \ \text{and} \ \mathbf{x}'_{(1)ijk,l}, \quad l = 1, 2, \ldots, c-1,$$

respectively. From (2.51) it follows that

$$y_{ijk} = \mathbf{X}_{(f)ijk}\boldsymbol{\beta} + \mathbf{X}_{(3)ijk}\mathbf{v}_i + \mathbf{X}_{(2)ijk}\mathbf{u}_{ij} + \mathbf{X}_{(1)ijk}\mathbf{e}_{ijk} \quad (2.52)$$
$$k = 1, 2, \ldots, s.$$

The set of regression equations given by (2.52) can be written as

$$\mathbf{y}_{ij} = \mathbf{X}_{(f)ij}\boldsymbol{\beta} + \mathbf{X}_{(3)ij}\mathbf{v}_i + \mathbf{X}^*_{(2)ij}\mathbf{u}^*_{ij}$$

where

$$\mathbf{X}^*_{(2)ij} = \begin{bmatrix} \mathbf{X}_{(2)ij1} & \mathbf{X}_{(1)ij1} & 0 & \cdots & 0 \\ \mathbf{X}_{(2)ij2} & 0 & \mathbf{X}_{(1)ij2} & \cdots & 0 \\ \vdots & \vdots & \vdots & \ddots & \vdots \\ \mathbf{X}_{(2)ijs} & 0 & 0 & \cdots & \mathbf{X}_{(1)ijs} \end{bmatrix}, \quad (2.53)$$

and

$$\mathbf{u}_{ij}^* = \begin{bmatrix} \mathbf{u}_{ij} \\ \mathbf{e}_{ij1} \\ \mathbf{e}_{ij2} \\ \vdots \\ \mathbf{e}_{ijs} \end{bmatrix},$$

where s is the maximum number of subpopulations.

Note that

$$\mathbf{X}_{(2)ij}^* = \begin{bmatrix} \mathbf{X}_{(2)ij} & \mathbf{X}_{(1)ij} \end{bmatrix},$$

where (cf. (2.46))

$$\mathbf{X}_{(1)ij} = \mathbf{D}_{ij}.$$

Suppose that for a given (i,j)-combination only $s_j < s$ subpopulations are present. In this case $\mathbf{X}_{(2)ij}$ has $(c-1) \times s_j$ rows, but the number of columns remains $m + (c-1) \times s$, which is the dimension of the random coefficient vector $\mathbf{u}_{ij}^*$.

The covariance of $\mathbf{u}_{ij}^*$ can then be written as

$$\boldsymbol{\Phi}_{(2)}^* = \mathrm{Cov}\left(\mathbf{u}_{ij}^*, \mathbf{u}_{ij}^{*\prime}\right) = \begin{bmatrix} \boldsymbol{\Phi}_{(2)} & 0 & \cdots & 0 \\ 0 & \boldsymbol{\Phi}_{(1)11} & \cdots & 0 \\ \vdots & \vdots & \ddots & \vdots \\ 0 & 0 & \cdots & \boldsymbol{\Phi}_{(1)ss} \end{bmatrix}. \quad (2.54)$$

2.3 Multilevel Input Files

Although users of the Windows version of the LISREL program have the option to construct the required command files through the dialog-box interface, we will focus on the use of the multilevel syntax to write a proper input file directly in a text editor, which is the way the program works on each platform that it is available for.[2]

Some analysis specifications cannot even be produced through the dialog screens, in which case the editing of the syntax file becomes necessary.

2.3.1 Data file

All the multilevel examples that are included with the program and discussed starting on page 70, use a PRELIS system file or PSF file as starting point. For example, the very first example contains the following command in the input file:

```
SY=K:\LISREL83\MLEVELEX\MOUSE.PSF;
```

The data are in a system file (indicated with the command SY=) with the name MOUSE.PSF in a certain directory on the system.

Here is an example of a simple PRELIS file that creates such a system file (file EX2.PR2):

```
EXAMPLE 2: ATTITUDES OF MORALITY AND EQUALITY
DA NI=8 NO=200 MI=0 TR=PA
LA
HUMRGHTS EQUALCON RACEPROB EQUALVAL EUTHANAS CRIMEPUN CONSCOBJ GUILT
RA FI=DATA.EX2
OU RA=DATA.PSF
```

After running PRELIS with this input file, a PRELIS system file DATA.PSF is created in the default directory.

We recommend that the user follows the same path and concentrates first on the data step using all the features that PRELIS offers. Once the data are in shape, create a PSF file and concentrate on the data analysis.

---

[2]SSI publishes a separate guide that includes the Windows dialog-box interface for multilevel analysis: *Interactive LISREL*.

2.3.2 Syntax file

This section provides an overview of the syntax conventions for multilevel analysis.

Syntax overview

The basic structure of the analysis input file is as given in Table 2.1, and the required commands are indicated. Turn to the page given in the last column for a detailed description of the particular command.

Table 2.1 Basic Structure of the Multilevel Input File

Command	Description	Required?	Page
OPTIONS	[list of options] ;	Yes	42
IDn =	[name of variable identifying level-n units] ;	Yes	46
RANDOMn =	[names of variables random on level n of the model] ;	Yes	47
RESPONSE =	[name(s) of response variables(s)] ;	Yes	48
FIXED =	[names of variables included as fixed effects in the model] ;	Yes	49
COVnPAT =	[pattern for level-n random coefficient covariance matrix] ;	No	50
COVnVAL =	[starting values for level-n random coefficient covariance matrix] ;	No	55
FIXVAL =	[starting values for fixed effect parameters] ;	No	56
CONTRAST =	[name of contrast file] ;	No	57
MISSING_DAT =	[integer value] ;	No	59
MISSING_DEP =	[integer value] ;	No	60
SUBPOP =	[names of variables to be used to construct subpopulations] ;	No	61
WEIGHT1 =	[name of level-1 weight variable] ;	No	62
TITLE =	[a descriptive title for the analysis] ;	No	62

Guidelines for constructing or changing the input file

When the input file is constructed or edited through a text editor, the following guidelines should be kept in mind:

- All commands start with a command name and conclude with a semi-colon.
- Note that the maximum length of a command in the input file is 80 characters.
- There is no specific required order in which commands have to be given, with the exception that the OPTIONS command must always be the first command in the input file.
- Blank lines between commands are allowed.
- The input file may be given in either upper or lower case.
- A keyword can be a mixture of upper and lower case letters.

OPTIONS

Required command

Each input file for a multilevel analysis starts with an OPTIONS command. The options are used to control the estimation procedure and the amount of output to be supplied at convergence of the iterative procedure.

The list of options may be given in any order between the command name and the semi-colon signaling the conclusion of this command. If the default values of the options are deemed sufficient, the corresponding options may also be omitted.

The options that may be used with the OPTIONS command and their default values are summarized below.

OLS

OLS estimates of the fixed effects are calculated as a first step of the iterative procedure unless otherwise specified. The OLS keyword is used to indicate whether the OLS estimates are to be calculated during the first iteration. Valid values are YES and NONE.

2.3 MULTILEVEL INPUT FILES

If NONE is specified, no OLS estimates will be calculated during the first iteration. The NONE value is used in combination with the optional FIXVAL command (see page 56) to allow the user to provide a set of initial values for the fixed coefficients in the model. The default value for this keyword is YES, indicating calculation of OLS estimates during iteration 1.

CONVERGENCE

A test for convergence is made at the end of each iteration. The difference in the numerical value between successive values of the estimated parameters is compared to a convergence criterion of the form 10^{-x}, where x is an integer with possible values $1, 2, \ldots$. If the difference between successive values is smaller than the convergence criterion, convergence is said to have been reached. The default convergence criterion is 0.001 (or 10^{-2}). In order to use a different value, for example, 0.0001 as convergence criterion, the specification CONVERGE = 0.0001 must be included on the OPTIONS command.

MAXITER

The keyword MAXITER is used to indicate the maximum number of iterations to be performed. The default number of iterations is 10, which should be sufficient for convergence to be reached in most cases. If, however, a more stringent convergence criterion is used or previous experience with a particular data set indicates slow convergence, this keyword may be used to increase the maximum number of iterations. If, on the other hand, the user wishes to obtain only the OLS estimates calculated in the first iteration, MAXITER should be set equal to 1.

OUTPUT

The OUTPUT keyword determines the amount of output required. Valid specifications are:

STANDARD	The default output only, as described below
BAYES	The default output and empirical Bayes estimates
RESIDUAL	The default output and residuals
TABLES	The default output and frequency tables
ALL	The default output, Bayes estimates, residuals, and tables

OUTPUT=STANDARD (Default output)

The following information is written to the default output file:

1. Input specifications as supplied by the user in the input file.
2. A summary of the hierarchical structure of the raw data.
3. Details of the iterative procedure. For each iteration, these details include the estimates, their standard errors, z-values, and probabilities of exceeding those limits.
4. The covariance and correlation matrices of the random coefficients on the different levels of the model are also given.
5. The value of $-2\ln L$ (likelihood function) at each iteration
6. The computation time for completion of the iterative procedure and writing of required results to the output file.

See also the section on the format of the default output file (p. 63) for a discussion of the default output file.

OUTPUT=BAYES (Empirical Bayes estimates)

If OUTPUT=BAYES is specified, 1 to 6 listed above are written to the output file. Two additional output files are created (in the case of a level-3 model).

The empirical Bayes estimates on levels 2 and 3 of the model are calculated and, along with their variance and relevant variable codes, are written to the files *.BA2 and *.BA3, where these filenames refer to the second and third level of the hierarchy, respectively.

See also the section on the format of the empirical Bayes estimate output file(s) (p. 68) for a discussion of the EB estimate output file obtained for the mice data.

OUTPUT=RESIDUAL (Residuals)

If OUTPUT=RESIDUAL is specified, items 1 to 6 listed above are written to the output file. An additional file, *.RES, is created, containing the residuals at convergence. The following information is given:

2.3 MULTILEVEL INPUT FILES

- the residuals $(y_{ijk} - \mathbf{x}'_{(f)ijk} * \hat{\boldsymbol{\beta}})$,
- expected value $(\tilde{y}_{ijk})$, and
- observed value (y_{ijk}) for each observation in the raw data set.

See also the section on the format of the residual output file (p. 69) for a discussion of the residual file obtained for the analysis of the mice data.

OUTPUT=TABLES (Frequency tables)

If OUTPUT=TABLES is specified, items 1 to 6 listed above are written to the output file. An additional file, *.FRQ, is created, containing a frequency table for each combination of level-3 and level-2 units. These frequency tables are constructed according to the subpopulations specified (see page 61).

Note that frequency tables can be generated only when the analysis is based on categorical outcome variables.

OUTPUT=ALL (All output)

All of the above files are created.

LINK

The LINK keyword may be used when a multilevel analysis with categorical outcome variable(s) is rrequested. At present, there are two specifications possible.

The default, LINK=0, specifies that the logit link function will be used. LINK=1 requests use of the cumulative logit link function.

ADD

The ADD keyword is used to indicate the value to be added to each cell of the frequency tables generated when a multilevel analysis with categorical outcome variable(s) is rrequested. The value specified should be a real number between zero and one, for example ADD=0.50.

The default value for this keyword is 0.25.

Examples of OPTIONS commands

(a) OPTIONS ;

By using this form of the OPTIONS command, the procedure to be used is IGLS and the convergence criterion is 0.001. A maximum number of 10 iterations will be carried out and partial output will be written to either the user specified output file or the default output file *.OUT. OLS estimates are calculated during the first iteration.

(b) OPTIONS MAXITER=5 OUTPUT=ALL;

This command specifies the use of a convergence criterion of 0.001 with the use of OLS estimation in the first iteration. A maximum of 5 iterations is allowed and the creation of all additional output files is requested.

(c) OPTIONS OLS=NONE MAXITER=20 OUTPUT=ALL CONVERGE=0.0001;

Use of this command will exclude the calculation of the OLS estimates during the first iteration. Subsequent iterations will be performed using the IGLS procedure. The convergence criterion is 10^{-4} and the maximum number of iterations 20, indicating that the iterative procedure will terminate according to these two criteria. If convergence is not reached after 20 iterations, the procedure will be terminated. In this case, lack of convergence will be noted in the default output file. Otherwise, the procedure will stop when convergence according to the specified criterion occurs before the maximum number of iterations. All additional output files as described under the OUTPUT keyword above will be created on the hard disk.

IDn

Required command

The ID commands are used to indicate the variable(s) identifying the units on the different levels of the hierarchy.

If the model specified by the user is a level-2 model, the commands ID1 and ID2 are required. Likewise, if a level-3 model is to be considered, the ID1, ID2, and ID3 commands are required in the input file. The exceptions

2.3 MULTILEVEL INPUT FILES

to this rule are in the case of a multivariate model and in the case of a model with no random component on level 1 of the model, where the ID1 command may be omitted.

Variables listed in the ID commands must be included in the PRELIS system file (*.PSF file). The spelling and case in which they are given need to correspond to that given in the PSF file. The syntax of this command is:

IDn = [variable name identifiying level-n units] ;

Examples of IDn commands

(a) If the raw data file contains information on the test scores, age, and gender of pupils belonging to classes within schools, and the variables school, class, pupil, age, gender, and result are contained in the PSF file, the following ID commands may be used to identify the levels of the hierarchical structure:

ID3=school ;

ID2=class ;

ID1=pupil;

(b) If the variables iden1, iden2, iden3, Y1, Y2, and X1 to X4 are in the PSF file, IDEN1 to IDEN3 may be used as shown below to identify the levels of the hierarchy.

ID3=iden3;

ID2=iden2;

ID1=iden1;

RANDOMn

Required command

One random command of the form

RANDOMn = [list of variable names] ;

is allowed for each level of the hierarchy. The RANDOM command is used to identify those coefficients that are allowed to vary randomly over a given level of the hierarchy.

Variables listed here must be included in the PSF file. The spelling and case in which they are given need to correspond to that given in the file.

Example of RANDOM commands

RANDOM3= X1:X4 ;

RANDOM2 = X2 X1;

RANDOM1 = X3;

From this hypothetical example, it can be seen that:

1. The random variables may be listed in any order.
2. Any or all of the possible predictors may be used at any level of the model.

As in the case of the ID commands, the RANDOM1 command may be omitted in the case of a multivariate model or if a model with no random component on level 1 of the hierarchy is to be fitted. Thus, in the case of a multivariate model, the following set of commands may be used:

```
ID3=iden3;
ID2=iden2;
RANDOM3= X1:X4 ;
RANDOM2= X3:X4 ;
```

It is possible to place constraints on elements of the random coefficient covariance matrices. Information on the constraints permitted and on the provision of initial values for elements of these matrices will be discussed when the COVnPAT and COVnVAL commands, which are optional, are considered.

RESPONSE

Required command

The RESPONSE command is of the form:

RESPONSE = [response variable(s)] ;

This command contains information on the response variable(s) to be used in the analysis.

In the case of a multivariate model, more than one response variable may be listed in the RESPONSE command. Spelling, etc., of the names of the response variables must once again be the same as that used in the PSF file and may be entered in any order.

Example of RESPONSE commands

(a) If the hypothetical example discussed previously is to be analyzed as a general level-3 model with Y1 as response variable, the relevant RESPONSE command will be:

RESPONSE = Y1;

(b) If, however, the same data is to be analyzed with a multivariate model with both Y1 and Y2 as response variables, the relevant RESPONSE command is either

RESPONSE = Y1 Y2 ;

or

RESPONSE = Y2 Y1 ;

FIXED

Required command

The FIXED command is used to identify the fixed effects for the model to be analyzed.

The syntax is:

FIXED = [list of variables names to be included as fixed effects] ;

The fixed effects may be all of the predictor variables contained in the raw data file or any subset of these predictors and may be specified in any order. Spelling of the names, however, must correspond to the spelling used in the PSF file.

If a covariate is included in the analysis, this should be reflected in the FIXED command. The format in which the covariate should be entered is:

FIXED = [var1] [...] [var$n$] [covariate*var1] [...] [covariate*var$n$];

Note that the covariate can be used in combination with any of the predictors listed in the FIXED command.

Examples of FIXED commands

(a) For the educational example, any one of the following FIXED commands is permissible:

FIXED = AGE ;

FIXED = GENDER ;

FIXED = GENDER AGE ;

(b) For the second of our previous examples, any of the following FIXED commands may be used:

FIXED X1:X2 ;

FIXED X1 X4 X3 X2 ;

or any other similar command.

(c) If the variable X3 is to be included as covariate in the first command in (b), the appropriate FIXED command is:

FIXED X1 X2 X3*X1 X3*X2;

Initial estimates for the fixed effects may be provided by the user. This is done through use of the optional FIXVAL command (discussed on p. 56).

COVnPAT

Optional command

The COVnPAT commands are used to place constraints on the covariance matrices of random coefficients on the different levels of the model.

The syntax for a COVnPAT command is:

COVnPAT= [options] ;

One COVnPAT command is allowed for each level of the hierarchy. If, for instance, a level-3 model with random components on all three levels of

2.3 MULTILEVEL INPUT FILES

the hierarchy is to be fitted, up to three COVnPAT commands may be included in the input file.

Options that may be used with this command are:

COVnPAT = DIAG;

In this case, the covariance matrix of random parameters on level n (where $n = 2$ or 3) of the model will be constrained to be a diagonal matrix of the form

$$\Phi_{(n)} = \begin{bmatrix} \phi_{(n)11} & 0 & \cdots & 0 \\ 0 & \phi_{(n)22} & \cdots & 0 \\ \vdots & \vdots & \ddots & \vdots \\ 0 & 0 & \cdots & \phi_{(n)pp} \end{bmatrix},$$

where p is the number of random coefficients on level n of the hierarchy.

COVnPAT = TOEPLITZ;

The covariance matrix on level n (where $n = 2$ or 3) will, in this case, be constrained to be of the form of a so-called Toeplitz matrix, *i.e.*

$$\Phi_{(n)} = \begin{bmatrix} \gamma_0 & \gamma_1 & \gamma_2 & \cdots & \gamma_{p-1} \\ \gamma_1 & \gamma_0 & \gamma_1 & \cdots & \gamma_{p-2} \\ \gamma_2 & \gamma_1 & \gamma_0 & \cdots & \gamma_{p-3} \\ \vdots & \vdots & \vdots & \ddots & \vdots \\ \gamma_{p-1} & \gamma_{p-2} & \gamma_{p-3} & \cdots & \gamma_0 \end{bmatrix}.$$

This option may also be invoked by using only the first four letters of the required structure, *i.e.*, TOEP.

COVnPAT = INTRA;

When the INTRA option is used, the covariance matrix of random parameters on level n (where $n = 2$ or 3) is constrained to have an intra-class structure, *i.e.*

$$\Phi_{(n)} = \begin{bmatrix} \alpha & \beta & \beta & \cdots & \beta \\ \beta & \alpha & \beta & \cdots & \beta \\ \beta & \beta & \alpha & \cdots & \beta \\ \vdots & \vdots & \vdots & \ddots & \vdots \\ \beta & \beta & \beta & \cdots & \alpha \end{bmatrix}.$$

This command may also be used with the abbreviation INTR as option.

COVnPAT = MA1;

In order to constrain the covariance matrix on level n to be similar to that of a time series process of order MA1, the MA1 option may be used. The form of the covariance matrix will then be

$$\Phi_{(n)} = \begin{bmatrix} \gamma & \beta & 0 & \cdots & 0 \\ \beta & \gamma & \beta & \cdots & 0 \\ 0 & \beta & \gamma & \cdots & 0 \\ \vdots & \vdots & \vdots & \ddots & \vdots \\ 0 & 0 & 0 & \cdots & \gamma \end{bmatrix}.$$

COVnPAT = LOGIT;

In the case of categorical data analysis it may be necessary to constrain the elements of the covariance matrix on level n to be of the following form

$$\Phi_{(n)} = \begin{bmatrix} \Psi_{11} & 0 & \cdots & 0 \\ 0 & \Psi_{22} & \cdots & 0 \\ \vdots & \vdots & \ddots & \vdots \\ 0 & 0 & \cdots & \Psi_{ss} \end{bmatrix},$$

where, for example in the case of 4 response categories, the submatrix Ψ_{ii} is of the form

$$\Psi_{ii} = \begin{bmatrix} \frac{1}{\pi_{i1}} + \frac{1}{\pi_{i4}} & \frac{1}{\pi_{i4}} & \frac{1}{\pi_{i4}} \\ \frac{1}{\pi_{i4}} & \frac{1}{\pi_{i2}} + \frac{1}{\pi_{i4}} & \frac{1}{\pi_{i4}} \\ \frac{1}{\pi_{i4}} & \frac{1}{\pi_{i4}} & \frac{1}{\pi_{i3}} + \frac{1}{\pi_{i4}} \end{bmatrix},$$

2.3 MULTILEVEL INPUT FILES

For each subpopulation i, π_{ij} denotes the probability of the j-th response ($j = 1, 2, 3, 4$) occurring. The subpopulations are defined using the optional SUBPOP command.

COVnPAT = [as specified by user] ;

If it is deemed necessary to constrain the elements of the covariance matrix to be of a form other than those discussed above, the user may specify this required structure through use of the COVnPAT command. This can be done by entering a lower-triangular matrix with the required structure, using the COVnPAT command. If, for example, the covariance matrix for the RANDOMn command

RANDOMn = X1 X2 X3 X4 ;

is to be constrained, this can be done by keeping in mind that the lower triangular elements of the covariance matrix are numbered row-wise as shown below.

```
1
2  3
4  5  6
7  8  9  10;
```

The elements to be fixed are then replaced with '0.' If, for example, the matrix is constrained to be diagonal, the command to be used is:

```
COVnPAT = 1
          0  3
          0  0  6
          0  0  0  10;
```

The structure as specified indicates that there are four **parameters to be estimated** (*i.e.*, numbers 1, 3, 6, and 10, corresponding to the variances) and five fixed parameters (corresponding to the covariances), indicated with 0. The values to which the fixed parameters are to be set equal to can be supplied using the COVnVAL command. If the COVnVAL command is omitted, the fixed parameters will be constrained to be equal to zero, as the initial structure of all covariance matrices are assumed to be diagonal at the start of the iterative procedure.

In the case of an MA1 process, for example, the command will be:

```
COVnPAT = 1
          2  1
          0  2  1
          0  0  2  1;
```

From this structure, it follows that there are only two parameters to be estimated (numbers 1 and 2) while all other parameters are constrained to be equal to zero, unless otherwise specified using the COVnVAL command.

It is permissible to constrain diagonal elements of the level-n covariance matrix to be equal to 0 through the use of the COVnVAL command.

The following commands, for example, are permissible

```
COVnPAT = 1
          2  0
          3  2  0
          0  0  2  0;

COVnPAT = 0
          2  0
          3  2  0
          0  0  2  0;
```

Note that zero values indicate that the corresponding elements remain fixed at the values specified in the COVnVAL commands.

In conclusion, the user should note that:

(a) No line of input may exceed 80 characters. Thus, if a large COVnPAT matrix is entered, care should be taken that no row of this matrix exceeds this limit. If a row of the matrix is too long, it may simply be continued on the next line of the input file.

(b) If elements of the covariance matrix to be estimated are constrained to be equal in value, the number assigned to those elements must be the same.

(c) As with all other commands in the input file, the command should end with a semi-colon that may be placed directly after the last element of the matrix as specified or on the next line of the input file.

2.3 MULTILEVEL INPUT FILES

(d) The matrix specified by the user must have the same number of elements as implied by the RANDOMn command. That is, if there are p variables listed in the RANDOMn command, a total number of $\frac{1}{2}p(p+1)$ elements must be entered by the user.

(e) In order to assign initial values to elements of the covariance matrix at level n or to set fixed elements of the matrix to user specified values, the COVnPAT command should be used in conjunction with the COVnVAL command.

COVnVAL

Optional command

COVnVAL commands may be used to provide either initial values for elements of the covariance matrix on level n of the model or to provide values for elements fixed through the use of options of the COVnPAT command.

The syntax of these commands is:

COVnVAL = [as specified by user] ;

One COVnVAL command is allowed for each level of the hierarchy. If, for instance, a level-3 model with random coefficients on all three levels of the hierarchy is to be fitted, up to three COVnVAL commands may be included in the input file.

The values to be used for the elements of the covariance matrix must be entered in the form of a lower-triangular matrix. The number of values entered must be the same as the number of elements implied by the relevant RANDOMn command. If there are p variables listed in the RANDOMn command, $\frac{1}{2}p(p+1)$ values must be entered by the user. If a large number of values are entered, a row of the lower-triangular matrix may be continued on the next line of the input file if the number of characters in that row of the matrix exceeds 80 characters. The command must end with a semicolon, which may be entered on the last line of the values given or on the next line of the input file.

Examples of COVnVAL commands

(a) Continuing with the example used to illustrate the use of the COVnPAT command to obtain a user specified covariance structure, the following command illustrates how the user may provide values for the elements of the covariance matrix (n).

```
COVnVAL = 1.00
         0.32  0.85
         0.63  0.62  0.78
         0.19  0.00  0.25  0.99;
```

If an accompanying COVnPAT command is not used, these values will function as starting values for the level-n covariance matrix. When good starting values for the elements of this covariance matrix are known, the use of the command as shown above together with the use of the option OLS=NONE on the OPTIONS command will likely decrease the number of iterations required to obtain convergence.

(b) When the command

COVnPAT = DIAG;

is used together with the command given in (a), the values specified on the diagonal of the lower-triangular matrix in (a) will be used as initial values for these parameters which are to be estimated. The off-diagonal elements of the covariance matrix will then be constrained to be equal to the values of off-diagonal elements of the matrix in (a).

FIXVAL

Optional command

It is also possible to provide initial values for the fixed parameters in the model to be analyzed. This may be achieved through the use of the FIXVAL command, which allows the user to provide starting values for these parameters. The syntax of this command is:

FIXVAL = [as specified by user] ;

2.3 MULTILEVEL INPUT FILES

The number of values entered by the user using this command must be equal to the number of fixed parameters to be estimated. There is no specific format in which the values have to be entered. The following three input styles

```
FIXVAL = 0.151 0.355 0.654;

FIXVAL = 0.151
         0.355
         0.654;
```

and

```
FIXVAL = 0.151
         0.355
         0.654
;
```

are all permissible. If the first of these commands is used and the number of characters in the user specified string exceeds 80 characters, the next line of the input file should be used.

The use of the FIXVAL command and the OLS=NONE option of the OPTIONS command may be particularly effective when good starting values of these parameters are available.

CONTRAST

Optional command

The CONTRAST command is used to specify the path to an optional additional input file containing information on any contrast(s) between the fixed effects in the model to be tested. The syntax of this command is similar to that of the PRELIS DA command in that the contrast file may be located in another directory or even subdirectory. Thus, both

```
CONTRAST = MLEVEL.CTR ;
```

and

```
CONTRAST= C:\MLEVEL\EXAMPLES\MLEVEL.CTR ;
```

are valid examples of the CONTRAST command. Information contained in the contrast file must adhere to certain specifications, as illustrated with the following examples.

Examples

Suppose that there are six fixed effects in a particular model: INTERCEPT, GENDER, MATHS, READING, SCIENCE, and WRITING.

If, for example, we wish to test

$$H_0: \quad \beta_{\text{READING}} - \beta_{\text{WRITING}} = 0$$
$$\beta_{\text{MATHS}} - \beta_{\text{SCIENCE}} = 0 \, ,$$

we can test this by specifying

$$H_0: \quad \mathbf{C}\boldsymbol{\beta} = \mathbf{0} \, ,$$

where

$$\mathbf{C} = \begin{bmatrix} 0 & 0 & 0 & 1 & 0 & -1 \\ 0 & 0 & 1 & 0 & -1 & 0 \end{bmatrix}$$

and

$$\boldsymbol{\beta}' = \begin{bmatrix} \beta_{\text{INTERCEPT}} & \beta_{\text{GENDER}} & \beta_{\text{MATHS}} & \beta_{\text{READING}} & \beta_{\text{SCIENCE}} & \beta_{\text{WRITING}} \end{bmatrix}$$

Note that each row of $\mathbf{C}$ has six elements, corresponding to the six fixed effects. Since the fourth element in the first row equals 1, this denotes β_{READING} while the sixth element denotes β_{WRITING}.

The contrast file will have the following form

2.3 MULTILEVEL INPUT FILES

```
2
0  0  0  1  0  -1
0  0  1  0  -1  0
```

The first row indicates the number of contrasts and the second and third rows the actual contrasts to be tested.

If the contrast file is specified as

```
1
0  0  0  1  0  -1
1
0  0  0  0  -1  0,
```

two separate contrast tests are performed as opposed to a simultaneous test for two contrasts.

MISSING_DAT

Optional command

There are two optional commands, which may be used when missing data is present in the raw data file. These commands are the MISSING_DAT and MISSING_DEP commands.

The MISSING_DAT command allows the user to specify an integer value, which will represent a missing value on any of the variables used in the analysis. The syntax of this command is

MISSING_DAT = [integer value] ;

Any positive or negative integer may be used. Only one value is allowed in this command. All records with data values equal to the code specified in this command will subsequently be removed from the analysis.

Examples of MISSING_DAT commands

Valid examples of the use of this command includes the following:

```
MISSING_DAT = 99 ;
MISSING_DAT = -998 ;
MISSING_DAT = 0 ;
```

Note that this command may also be used in conjunction with the MISSING_DEP command.

MISSING_DEP

Optional command

The MISSING_DEP command may be used to specify a code assigned to missing values on the response variables only. The syntax of this command is

MISSING_DEP = [integer value] ;

The same rules for values, which may be used with the MISSING_DAT command, applies to the MISSING_DEP command. The consequence of using the MISSING_DEP command is that only records with response variable values equal to the code assigned through the MISSING_DEP command will be removed from the analysis.

The MISSING_DEP command is recommended for use in the case of multivariate analysis. If only one of the response variables to be used in the multivariate analysis has a missing response, only that particular response will be considered missing while the remaining responses will still be used.

Example of a MISSING_DEP command

Consider the observations

Response variables				Predictor variables			
4.0	5.3	1.7	99	1	10	14.5	999
3.2	4.4	99	7.7	3	12	13.7	53.2

and the command

```
MISSING_DEP=99 ;
```

If the code 99 is identified as the code for missing data values on the dependent variables, this will imply that the analysis of this record will use the first three response values and disregard the fourth one in the case of the first observation. The third response variable will, however, be removed where the second observation is concerned.

2.3 MULTILEVEL INPUT FILES

If the code 999 is specified as the code for missing data values on all the variables included in the analysis, however, the whole first record as given above will be deleted from the data set to be analyzed. The second observation will be retained with the exception of the third response variable value.

This is accomplished by using both the MISSING_DEP and MISSING_DAT commands:

```
MISSING_DEP = 99 ;
MISSING_DAT = 999;
```

Note that if only the MISSING_DEP command is used for the two observations given above, the value of 999 for the last predictor variable on the first observation will be considered valid data and will be used as such in the analysis.

SUBPOP

Optional command

When categorical data are to be analyzed, subpopulations may be created through use of the SUBPOP command.

The syntax of this command is:

SUBPOP = [names of variables used to create subpopulations] ;

Use of, for example, two variables with two response categories each will lead to the creation of four subpopulations, *i.e.*, a two-way table for these two variables. Counts of responses to a particular question for these subpopulations may then be used as the response variable in a categorical data analysis.

Example of a SUBPOP command

Consider the two variables GENDER and AGE. If GENDER has two possible outcomes, for example, 1 = Male and 2 = Female and AGE has three outcomes, for example, 1 = less than 20 years old, 2 = 21–40 years old, and 3 = 41+ years old, the use of the SUBPOP command

SUBPOP = GENDER AGE;

will induce the creation of six subpopulations for the combination (GENDER;AGE), namely:

(GENDER=1;AGE=1)   (GENDER=1;AGE=2)   (GENDER=1;AGE=3)
(GENDER=2;AGE=1)   (GENDER=2;AGE=2)   (GENDER=2;AGE=3)

WEIGHT1

Optional command

The WEIGHT1 command is only available for the analysis of categorical outcome variables. The inverse of the selection probability of level-1 units should be used as the weight variable.

The syntax of this command is:

WEIGHT1 = [name of level-1 weight variable] ;

TITLE

Optional command

The TITLE command allows the user to provide a description of the analysis to be performed. This command, like all commands excluding the OPTIONS command, can be placed anywhere in the input file. The maximum permissible length of this command is 70 characters. The syntax of this command is

TITLE = [title as provided by the user] ;

2.4 Multilevel Output Files

In this section, the list output and the optional additional output files are discussed.

Each successful analysis generates list output that is displayed on the screen for review and saved to a file with the extension OUT. The following is a reference overview of the different output sections that a multilevel analysis may generate. It uses a default output file from one of the examples (*Analysis of 2-level repeated measures data*, starting on p. 70), annotated with brief descriptions of the various parts.

There are three optional output files that may be saved by the program after each analysis. They are discussed in the sections *Empirical Bayes estimates: the output files *.BA2 and *.BA3* (see p. 68), and *Residuals: the output file *.RES* (see p. 69).

The data set used contains repeated measurements on 82 striped mice and was obtained from the Department of Zoology at the University of Pretoria, South Africa (see du Toit, 1979). A number of male and female mice were released in an outdoor enclosure with nest boxes and sufficient food and water. They were allowed to multiply freely. Occurrence of birth was recorded daily and newborn mice were weighed weekly, from the end of the second week after birth until physical maturity was reached. The data set consists of the weights of 42 male and 40 female mice. For male mice, 9 repeated weight measurements are available and for the female mice 8 repeated measurements. For a detailed analysis of this data, please see the example *Analysis of 2-level repeated measures data*, starting on p. 70.

The annotated output file for a level-2 analysis follows.

Analysis requested

The first section of the output file contains an echo of the input file used for analysis. It is concluded with a message indicating that no problems occurred during the processing of the input file. If an error had occurred, for instance if the OPTIONS command was missing or one of the keywords was misspelled, an appropriate error message will be displayed here.

```
The following lines were read from file K:\LISREL83\MLEVELEX\MOUSE4.PR2

OPTIONS OLS=YES CONVERGE=0.001000 MAXITER=10 OUTPUT=ALL ;
TITLE=Mouse data: variance decomposition;
SY=K:\LISREL83\MLEVELEX\MOUSE.PSF;
ID1=iden1;
ID2=iden2;
RESPONSE=weight;
FIXED=constant time timesq;
RANDOM1=constant;
RANDOM2=constant time timesq;
```

NO ERROR DIAGNOSTICS GENERATED AT MODEL SPECIFICATIONS STAGE

Data summary

The data summary section contains information on the number of units at different levels of the hierarchy. From the data summary given here, it can be concluded that there were eight or nine separate measurements for each level-2 unit, the level-2 units being the 82 mice.

```
                          +---------------+
                          | DATA SUMMARY  |
                          +---------------+

NUMBER OF LEVEL 2 UNITS :       82
NUMBER OF LEVEL 1 UNITS :      698

        N2  :     1      2      3      4      5      6      7      8
        N1  :     9      9      9      9      9      9      9      9

        N2  :     9     10     11     12     13     14     15     16
        N1  :     9      9      9      9      9      9      9      9

        N2  :    17     18     19     20     21     22     23     24
        N1  :     9      9      9      9      9      9      9      9

        N2  :    25     26     27     28     29     30     31     32
        N1  :     9      9      9      9      9      9      9      9

        N2  :    33     34     35     36     37     38     39     40
        N1  :     9      9      9      9      9      9      9      9

        N2  :    41     42     43     44     45     46     47     48
        N1  :     9      9      8      8      8      8      8      8

        N2  :    49     50     51     52     53     54     55     56
        N1  :     8      8      8      8      8      8      8      8
```

```
N2 :    57   58   59   60   61   62   63   64
N1 :     8    8    8    8    8    8    8    8

N2 :    65   66   67   68   69   70   71   72
N1 :     8    8    8    8    8    8    8    8

N2 :    73   74   75   76   77   78   79   80
N1 :     8    8    8    8    8    8    8    8

N2 :    81   82
N1 :     8    8
```

Fixed part of model

Estimates of the elements of the fixed coefficient vector at each iteration are given here. The standard errors, z-values, and probabilities of exceeding those limits are also reported.

Note that, before convergence is attained, the standard errors, z-values, and probabilities of exceeding are only an indication of the relevant values at convergence. In the output shown below, all the fixed effects are statistically significant.

```
Mouse data: variance decomposition

ITERATION NUMBER     5

                  +----------------------+
                  |  FIXED PART OF MODEL |
                  +----------------------+

------------------------------------------------------------------------
COEFFICIENTS        BETA-HAT      STD.ERR.       Z-VALUE       PR > |Z|
------------------------------------------------------------------------
constant             4.16213       0.45748        9.09788       0.00000
time                 6.90560       0.30980       22.29056       0.00000
timesq              -0.29629       0.02960      -10.01009       0.00000
```

Log-likelihood value

The log-likelihood value reported here is the value of $-2\ln L$ (likelihood function) evaluated at the parameter values during the relevant iteration.

Comparison of these values over iterations gives an indication of how stable the iterative procedure is. For a model that provides a good description of the data, this statistic should decrease from one iteration to the next.

It may also be used to evaluate the goodness of fit of different models for the same data. The difference between the $-2\ln L$ values for models based on the same data has a chi-square distribution. The degrees of freedom equal the difference in the number of parameters estimated in these models.

```
            +----------------------+
            |   -2 LOG-LIKELIHOOD  |
            +----------------------+

-2 LOG-LIKELIHOOD =    3400.93248789791
```

Random part of model

This section contains a list of estimates of all the free elements of the covariance matrices of the random parameters at the different levels of the hierarchy. The standard errors, z-values, and the probabilities of exceeding those limits ($\text{Prob}(Z > z \text{ or } Z < -z)$) are also given. Note that, before convergence is attained, the standard errors, z-values, and probabilities of exceeding are only an indication of the relevant values at convergence.

The coefficient CONSTANT, denoting the intercept term, has the largest variation over the level-2 units. From the output it is seen that all variances and covariances are significant.

```
            +----------------------+
            |  RANDOM PART OF MODEL |
            +----------------------+
```

LEVEL 2	TAU-HAT	STD.ERR.	Z-VALUE	PR > |Z|
constant/constant	11.72906	2.70296	4.33933	0.00001
time /constant	-5.86552	1.58228	-3.70699	0.00021
time /time	6.59372	1.23130	5.35509	0.00000
timesq /constant	0.38927	0.14067	2.76733	0.00565
timesq /time	-0.55909	0.11275	-4.95880	0.00000
timesq /timesq	0.05814	0.01124	5.17424	0.00000

2.4 MULTILEVEL OUTPUT FILES

```
------------------------------------------------------------------
LEVEL 1                       TAU-HAT    STD.ERR.    Z-VALUE    PR > |Z|
------------------------------------------------------------------
constant/constant             3.07863    0.20465     15.04328   0.00000
```

Random coefficient covariance and correlation matrices

In this section, the covariance and correlation matrices of the random coefficients in the model are given. These matrices include the elements that may have been considered fixed during the iterative procedure.

In certain situations, there may be differences between elements listed in the section of random parameter matrices and those given here. This will be the case if a covariance matrix was not positive definite. The matrix reported will be adjusted to be positive definite. This is accomplished by replacing the negative eigenvalues with a small positive number in the corresponding covariance matrix. If there are large differences between the elements reported in these two sections, this may indicate a misspecification of the model or the selection of an inappropriate covariance pattern.

Problems of this nature will also be reflected in the correlation matrix for the same level of the model as correlations close to either 1 or -1 in value. For this reason, the correlation matrices are provided as part of the output.

```
           LEVEL 2 COVARIANCE MATRIX

              constant      time       timesq

constant      11.72906
time          -5.86552     6.59372
timesq         0.38927    -0.55909     0.05814


           LEVEL 2 CORRELATION MATRIX

              constant      time       timesq

constant       1.0000
time          -0.6670      1.0000
timesq         0.4714     -0.9030      1.0000
```

```
        LEVEL 1 COVARIANCE MATRIX

                constant

constant     3.07863


        LEVEL 1 CORRELATION MATRIX

                constant

constant     1.0000
```

Technical details

This section is the last part of the output and gives the number of iterations performed as well as the total CPU time used for the iterative procedure.

```
CONVERGENCE REACHED IN   5 ITERATIONS
```

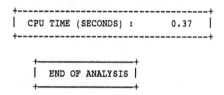

Empirical Bayes estimates: the output files *.BA2 and *.BA3

The model fitted in this example was a level-2 model. A description of the data and model are given on page 63.

For a two level model no *.BA3 file is created. The structure of a BA3 file for a 3-level model is similar to that of the file MOUSE4.BA2.

The contents of the output file MOUSE4.BA2 for mice 7 and 8 are as follows:

```
        7       1       1.5979         3.3066        constant
        7       2      -1.0797         0.73621       time
        7       3       0.25660        0.73356E-02   timesq
        8       1      -1.7937         3.3066        constant
        8       2       3.2809         0.73621       time
        8       3      -0.22528        0.73356E-02   timesq
```

2.4 MULTILEVEL OUTPUT FILES

The first column of the output file MOUSE4.BA2 indicates the level-2 unit, in this case a specific mouse. There are three fixed effects in the model fitted to the data and this is reflected in the second column which gives the number of the fixed effect for each mouse.

The third column contains the deviations of the empirical Bayes estimates from the estimated population parameters. For mouse 7, it follows from the results for the fixed part of the model in the default output file that

$$\hat{\beta}' = \begin{bmatrix} 4.16213 & 6.90560 & -0.29629 \end{bmatrix}.$$

The deviations from these parameters for mouse 7, for example, are 1.5979, -1.0797, and 0.25660, respectively. Hence, the empirical Bayes estimates of the vector of fixed parameters for mouse 7 is given by

$$\begin{aligned} 4.16213 + 1.5979 &= 5.76003 \\ 6.90560 - 1.0797 &= 5.8259 \\ -0.29629 + 0.25660 &= -0.03969 \end{aligned}$$

The fourth column of the MOUSE4.BA2 file gives the variances of the empirical Bayes estimates while the last column contains the name of the relevant fixed effect.

Residuals: the output file *.RES

This file contains one line of information for each of the observations from the raw data file that was used in the analysis. The first 4 lines of this file for the mouse data analysis are:

1	1	1	15.000	10.771	4.2286
2	1	2	17.000	16.788	0.21183
3	1	3	23.000	22.212	0.78768
4	1	4	24.000	27.044	-3.0439

The first three columns of this file consist of information identifying the observations. In the first column, the residuals are numbered sequentially, in the second column the level-2 ID number identifying the 82 mice and in the third column the time points at which measurements were made (*i.e.*, the level-1 ID number) are given.

The fourth column contains the observed values of the response variables for each observation and thus represents the vector y. This is followed by the expected values which represent the vector $X_{(f)} * \hat{\beta}$. Finally, the residuals are given, which are calculated as $y - X_{(f)} * \hat{\beta}$, being the difference between the observed and estimated values of the response variable.

2.5 Multilevel Examples

This section presents four examples[3] using multilevel analysis: An analysis of 2-level repeated measures data (see p. 70); a multivariate analysis of educational data (see p. 102); an analysis of air traffic control data (see p. 92); and an analysis of CPC Survey data (see p. 113). For a brief review of the various parts of the default output for multilevel analysis, see p. 63.

2.5.1 Analysis of 2-level repeated measures data

This example illustrates how multilevel modeling may be used to recognize explicitly the hierarchical structure of repeated measurement data.

Five models will be fitted and discussed:

- A variance decomposition model
- Modeling linear growth
- Modeling non-linear growth
- Introducing a covariate when modeling non-linear growth
- A model with complex variation at level 1 of the hierarchy

---

[3] Additional examples can be found on SSI"s website: www.ssicentral.com.

2.5 MULTILEVEL EXAMPLES

Description of the data

The data set used contains repeated measurements on 82 striped mice and was obtained from the Department of Zoology at the University of Pretoria, South Africa (see du Toit, 1979). A number of male and female mice were released in an outdoor enclosure with nest boxes and sufficient food and water. They were allowed to multiply freely. Occurrence of birth was recorded daily and newborn mice were weighed weekly, from the end of the second week after birth until physical maturity was reached. The data set consists of the weights of 42 male and 40 female mice. For male mice, 9 repeated weight measurements are available and for the female mice 8 repeated measurements.

The first 11 observations from this data set, contained in MOUSE.PSF, and the variable names to be used are shown below.

IDEN2	IDEN1	WEIGHT	CONSTANT	TIME	TIMESQ	GENDER
1.00	1.00	15.00	1.00	1.00	1.00	1.00
1.00	2.00	17.00	1.00	2.00	4.00	1.00
1.00	3.00	23.00	1.00	3.00	9.00	1.00
1.00	4.00	24.00	1.00	4.00	16.00	1.00
1.00	5.00	26.00	1.00	5.00	25.00	1.00
1.00	6.00	31.00	1.00	6.00	36.00	1.00
1.00	7.00	37.00	1.00	7.00	49.00	1.00
1.00	8.00	42.00	1.00	8.00	64.00	1.00
1.00	9.00	46.00	1.00	9.00	81.00	1.00
2.00	1.00	11.00	1.00	1.00	1.00	1.00
2.00	2.00	14.00	1.00	2.00	4.00	1.00

The response variable WEIGHT contains the weight measurements (in grams) for all mice at the different times of measurement. The explanatory variables which may be used are the time points at which measurements were made (TIME), the squared values of these time points (TIMESQ), and the gender of the mice (GENDER). It is also assumed that the growth of the mice during this period can be adequately described with a parabolic function.

A hierarchical level-2 structure is incorporated where the individual mice are the level-2 units. Unique numbers identifying the mice are contained in the variable IDEN2, which will be used as the level-2 ID for the analysis. The variable IDEN1 identifies the occasions on which measurements for a particular mouse were made and will be used as the level-1 ID. From the description of the data set as given above, it follows that there are 82 level-2 units, with either 8 or 9 measurements nested within each level-2 unit.

The variable CONSTANT consists of a column of 1s and will be used to estimate the intercept term in the model. It may also be used to estimate the amount in which the intercept varies over the different levels of the hierarchy, as illustrated in the following section.

Variance decomposition

The simplest multilevel model is equivalent to a one-way ANOVA with random effects. Although this model is not interesting in itself, it is useful as a preliminary step in a multilevel analysis as it provides important information about the outcome variability at each of the levels of the hierarchy. It may also function as a baseline with which more sophisticated models may be compared.

Let the subscript i denote the i-th level-2 unit, in this case the i-th mouse. The subscript j refers to the j-th weight measurement for the i-th mouse. Using this notation, the one-way ANOVA model can be written as:

$$y_{ij} = \text{CONSTANT}_{ij}\beta_0 + \text{CONSTANT}_{ij}u_{ij} + \text{CONSTANT}_{ij}e_{ij}$$

where u_{ij} denotes the random component on level 2 of the model. It is assumed that u_{ij} has an expected value of 0 and a variance of $\Phi_{(2)}$. The variance $\Phi_{(2)}$ may be interpreted as the 'between-group' variability. Likewise, it is assumed that e_{ij} is $N(0, \Phi_{(1)})$ distributed. Thus $\Phi_{(1)}$ may be interpreted as the 'within-group' variability.

This model is also known as a fully unconditional model (Bryk & Raudenbush, 1992), as no predictors are specified at either level of the hierarchy.

Leave all the options default, or simply cut and paste them from another input file. Type a title command for this analysis. In this example we use *Mouse data: Variance decomposition*. Next, specify the location of the PSF file.

2.5 MULTILEVEL EXAMPLES

Select the variable IDEN2 as the level-2 identification variable and IDEN1 as the level-1 identification variable. Select the variable WEIGHT as response variable and the variable CONSTANT as fixed variable.

This variable CONSTANT, representing the intercept term, is then also specified as the level-1 and level-2 variable associated with the error terms.

The complete syntax file MOUSE1.PR2 should look like this:

```
OPTIONS OLS=YES CONVERGE=0.001000 MAXITER=10 OUTPUT=STANDARD ;
TITLE=Mouse data: Variance decomposition;
SY=K:\LISREL83\MLEVELEX\MOUSE.PSF;
ID1=iden1;
ID2=iden2;
RESPONSE=weight;
FIXED=constant;
RANDOM1=constant;
RANDOM2=constant;
```

With the exception of the use of the optional TITLE command, this input file is the most basic one which can be used for the analysis of a level-2 model. Note that in the OPTIONS command the default values of the options MAXITER, CONVERGE, and OUTPUT are used (see the section on syntax on p. 41 for more detailed information).

Run PRELIS with this input file. Convergence is achieved after 3 iterations, and the details of the last iteration, as given in the output file, are given below.

```
                        +---------------+
                        | DATA SUMMARY  |
                        +---------------+

NUMBER OF LEVEL 2 UNITS :      82
NUMBER OF LEVEL 1 UNITS :     698

    N2 :     1      2      3      4      5      6      7      8
    N1 :     9      9      9      9      9      9      9      9

    N2 :     9     10     11     12     13     14     15     16
    N1 :     9      9      9      9      9      9      9      9

    N2 :    17     18     19     20     21     22     23     24
    N1 :     9      9      9      9      9      9      9      9
```

```
N2  :    25   26   27   28   29   30   31   32
N1  :     9    9    9    9    9    9    9    9

N2  :    33   34   35   36   37   38   39   40
N1  :     9    9    9    9    9    9    9    9

N2  :    41   42   43   44   45   46   47   48
N1  :     9    9    8    8    8    8    8    8

N2  :    49   50   51   52   53   54   55   56
N1  :     8    8    8    8    8    8    8    8

N2  :    57   58   59   60   61   62   63   64
N1  :     8    8    8    8    8    8    8    8

N2  :    65   66   67   68   69   70   71   72
N1  :     8    8    8    8    8    8    8    8

N2  :    73   74   75   76   77   78   79   80
N1  :     8    8    8    8    8    8    8    8

N2  :    81   82*
N1  :     8    8

ITERATION NUMBER      3
```

+------------------------+
| FIXED PART OF MODEL |
+------------------------+

COEFFICIENTS	BETA-HAT	STD.ERR.	Z-VALUE	PR > |Z|
constant	28.63410	0.57021	50.21634	0.00000

+------------------------+
| -2 LOG-LIKELIHOOD |
+------------------------+

-2 LOG-LIKELIHOOD =   5425.49001592990

+------------------------+
| RANDOM PART OF MODEL |
+------------------------+

LEVEL 2	TAU-HAT	STD.ERR.	Z-VALUE	PR > |Z|
constant/constant	11.32910	4.25185	2.66451	0.00771

2.5 MULTILEVEL EXAMPLES

```
------------------------------------------------------------------
LEVEL 1                    TAU-HAT      STD.ERR.    Z-VALUE   PR > |Z|
------------------------------------------------------------------
constant/constant         130.32083     7.42514     17.55130   0.00000

         LEVEL 2 COVARIANCE MATRIX

                 constant

constant    11.32910

         LEVEL 2 CORRELATION MATRIX

                 constant

constant    1.0000

         LEVEL 1 COVARIANCE MATRIX

                 constant

constant   130.32083

         LEVEL 1 CORRELATION MATRIX

                 constant

constant    1.0000
```

In the first part of the abbreviated output file shown here, the data summary for the hierarchical structure is given. The first 42 level-2 units are the male mice and the last 40 the female mice.

From the random part of the output it can be seen that the variation over measurements (level 1) is large and overwhelms the variation between the mice (level 2). The so-called intraclass correlation can be calculated as:

$$\hat{\rho} = \frac{\hat{\Phi}_{(2)}}{\hat{\Phi}_{(2)} + \hat{\Phi}_{(1)}} = \frac{11.32910}{11.32910 + 130.32083} = 0.0799$$

indicating that about 8 percent of the variance in weight measurements is between mice. The value of $-2\ln L$ (likelihood function) at convergence is 5425.4900.

Modeling linear growth

The variance decomposition model may now be extended by including the variable TIME as a fixed effect in the model. The model thus becomes

$$y_{ij} = \text{CONSTANT}_{ij}\beta_0 + \text{TIME}_{ij}\beta_1 + \text{CONSTANT}_{ij}u_{ij} + \text{CONSTANT}_{ij}e_{ij}$$

with the variable TIME used as predictor of the response measurements.

The only change to the input file MOUSE2.PR2 is in the FIXED command, which now becomes:

```
FIXED=CONSTANT TIME;
```

The details of the last iteration for this model are:

```
ITERATION NUMBER 3
                        +-----------------------+
                        | FIXED PART OF MODEL   |
                        +-----------------------+
```

COEFFICIENTS	BETA-HAT	STD.ERR.	Z-VALUE	PR > |Z|
constant	9.09586	0.60387	15.06258	0.00000
time	4.09218	0.06258	65.39108	0.00000

```
                        +-----------------------+
                        | -2 LOG-LIKELIHOOD     |
                        +-----------------------+

-2 LOG-LIKELIHOOD =     4137.57876020825

                        +-----------------------+
                        | RANDOM PART OF MODEL  |
                        +-----------------------+
```

LEVEL 2	TAU-HAT	STD.ERR.	Z-VALUE	PR > |Z|
constant/constant	20.69397	3.53655	5.85146	0.00000

LEVEL 1	TAU-HAT	STD.ERR.	Z-VALUE	PR > |Z|
constant/constant	16.46288	0.93806	17.54996	0.00000

2.5 MULTILEVEL EXAMPLES

```
            LEVEL 2 COVARIANCE MATRIX

                    constant

constant    20.69397


            LEVEL 2 CORRELATION MATRIX

                    constant

constant    1.0000


            LEVEL 1 COVARIANCE MATRIX

                    constant

constant    16.46288


            LEVEL 1 CORRELATION MATRIX

                    constant

constant    1.0000
```

- Both the fixed effects are highly significant, indicating significant variation in the intercepts and effect of time of measurement on the response variable over the different mice. An expected increase of 4.0922 grams in weight is expected for each increase of a week in TIME.
- The log-likelihood value for this model is 4137.5788, compared to the value of 5425.4900 for the fully unconditional model. This reduction in the log-likelihood value indicates considerable variation between mice in their linear growth rates and also that the model fitted here explains more of the variation in the data than the previous one.
- The random variation on level 2 of the model is higher and that on level-1 lower than in the fully unconditional model. The amount of variation in weight between mice is now calculated as

$$\hat{\rho} = \frac{\hat{\Phi}_{(2)}}{\hat{\Phi}_{(2)} + \hat{\Phi}_{(1)}} = \frac{20.69379}{20.69379 + 16.46288} = 0.5569 \;,$$

that is 56 percent.

□ From the reduction in the level-1 variance component it can be seen that the variable TIME accounts for a considerable part of the variance previously noted on this level.

It is expected that the linear growth rate may vary from mouse to mouse around its mean value, rather than be fixed. The random component on level 2 of the hierarchy is thus extended to (MOUSE3.PR2):

```
RANDOM2 = CONSTANT TIME ;
```

After running PRELIS on this input file to fit the model to the data, we find the output for this model (here abbreviated) in the MOUSE3.OUT file:

```
ITERATION NUMBER      4
```

+------------------------+
| FIXED PART OF MODEL |
+------------------------+

COEFFICIENTS	BETA-HAT	STD.ERR.	Z-VALUE	PR > |Z|
constant	9.20384	0.46702	19.70746	0.00000
time	4.05978	0.12358	32.85144	0.00000

+------------------------+
| -2 LOG-LIKELIHOOD |
+------------------------+

-2 LOG-LIKELIHOOD =   3873.66054782870

+------------------------+
| RANDOM PART OF MODEL |
+------------------------+

LEVEL 2	TAU-HAT	STD.ERR.	Z-VALUE	PR > |Z|
constant/constant	12.85320	2.80949	4.57492	0.00000
time /constant	-1.64430	0.59236	-2.77584	0.00551
time /time	1.07389	0.19582	5.48415	0.00000

LEVEL 1	TAU-HAT	STD.ERR.	Z-VALUE	PR > |Z|
constant/constant	8.90088	0.54469	16.34126	0.00000

2.5 MULTILEVEL EXAMPLES

```
              LEVEL 2 COVARIANCE MATRIX

                    constant       time

constant          12.85320
time              -1.64430      1.07389


              LEVEL 2 CORRELATION MATRIX

                    constant       time

constant           1.0000
time              -0.4426        1.0000


              LEVEL 1 COVARIANCE MATRIX

                    constant

constant           8.90088


              LEVEL 1 CORRELATION MATRIX

                    constant

constant           1.0000
```

- Once again a considerable reduction in the value of the function $-2\ln L$ is noted. The estimates for the fixed effects in the model stayed fairly constant. On level 2 of the model we see that all three elements of the covariance matrix of random parameters $\Phi_{(2)}$ are significant at a 5% level. The correlation between the CONSTANT term and the TIME term is given as -0.4426.

- The level-1 or error variance is further reduced to 8.90088. It can thus be concluded that the inclusion of the variable TIME significantly reduced the variation between measurements, *i.e.*, the level-1 units.

- The total variation on a particular level of the hierarchy may also be calculated. In this case, the total variance on level 2 is the variance of the sum of the two random coefficients associated with CONSTANT and TIME and may be written as:

$$\text{Var}\left(\text{CONSTANT}_{ij} + \text{TIME}_{ij}\,\Phi_{(2)\text{TIME,TIME}}\right)$$

$$= \Phi_{(2)\text{CONSTANT,CONSTANT}} + 2\Phi_{(2)\text{TIME,CONSTANT}}(\text{TIME}_{ij}) +$$

$$+ \Phi_{(2)\text{TIME,TIME}}(\text{TIME}_{ij})^2$$

$$= 12.85320 + 2(-1.64430)(\text{TIME}_{ij}) + 1.07389(\text{TIME}_{ij})^2$$

We can thus write the total variation at level 2 as a quadratic function of the variable TIME. A graph of this total variance against the nine time points is given in Figure 2.1.

❑ The increase in variance over time is to be expected with data of this nature. It could also be an indication that the assumption that a parabolic function can adequately describe this phase in the development of the mice may not be valid and that other functions should be considered.

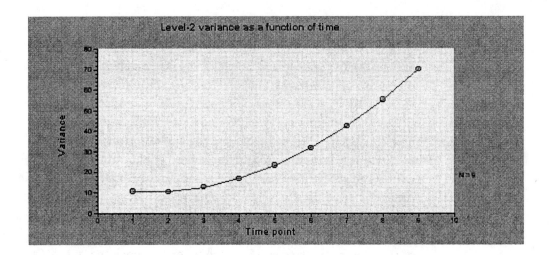

Figure 2.1   Between-mice variation as a function of the time points

2.5 MULTILEVEL EXAMPLES

Modeling non-linear growth

In data of this nature, it is unlikely that the increase in weight measurement will be linear for all mice over the time period concerned. A non-linear component may be introduced in the model discussed in the previous section by adding a quadratic term (the variable TIMESQ) to the model. The model previously given is thus extended to:

$$\begin{aligned} y_{ij} &= \text{CONSTANT TIME TIMESQ}_{ij}\beta + \text{CONSTANT TIME TIMESQ}_{ij}u_{ij} + \\ & \quad \text{CONSTANT}_{ij}e_{ij} \\ &= \text{CONSTANT}_{ij}\beta_0 + \text{TIME}_{ij}\beta_1 + \text{TIMESQ}_{ij}\beta_2 + \\ & \quad \text{CONSTANT}u_{0i} + \text{TIME}u_{1i} + \text{TIMESQ}u_{2i} + \\ & \quad \text{CONSTANT}_{ij}e_{ij} \end{aligned}$$

The addition of the variable TIMESQ in this case leads to the following changes in the FIXED and RANDOM2 commands contained in the input file.

```
FIXED = CONSTANT  TIME   TIMESQ ;
RANDOM2 = CONSTANT  TIME   TIMESQ ;
```

In order to obtain the empirical Bayes residuals for the level-2 models and the fitted values for each observation, the option OUTPUT=BAYES is added to the OPTIONS command.

The complete input file (MOUSE4.PR2) is now:

```
OPTIONS OLS=YES CONVERGE=0.001000 MAXITER=10 OUTPUT=ALL ;
TITLE=Mouse data: Modeling non-linear growth ;
SY=K:\LISREL83\MLEVELEX\MOUSE.PSF;
ID1=iden1;
ID2=iden2;
RESPONSE=weight;
FIXED=constant time timesq;
RANDOM1=constant;
RANDOM2=constant time timesq;
```

Convergence of the iterative procedure was reached in five iterations, producing the following output.

ITERATION NUMBER   5

```
            +----------------------+
            | FIXED PART OF MODEL  |
            +----------------------+
```

COEFFICIENTS	BETA-HAT	STD.ERR.	Z-VALUE	PR > |Z|
constant	4.16213	0.45748	9.09788	0.00000
time	6.90560	0.30980	22.29056	0.00000
timesq	-0.29629	0.02960	-10.01009	0.00000

```
            +----------------------+
            |  -2 LOG-LIKELIHOOD   |
            +----------------------+
```

-2 LOG-LIKELIHOOD =   3400.93248789791

```
            +----------------------+
            | RANDOM PART OF MODEL |
            +----------------------+
```

LEVEL 2	TAU-HAT	STD.ERR.	Z-VALUE	PR > |Z|
constant/constant	11.72906	2.70296	4.33933	0.00001
time /constant	-5.86552	1.58228	-3.70699	0.00021
time /time	6.59372	1.23130	5.35509	0.00000
timesq /constant	0.38927	0.14067	2.76733	0.00565
timesq /time	-0.55909	0.11275	-4.95880	0.00000
timesq /timesq	0.05814	0.01124	5.17424	0.00000

LEVEL 1	TAU-HAT	STD.ERR.	Z-VALUE	PR > |Z|
constant/constant	3.07863	0.20465	15.04328	0.00000

LEVEL 2 COVARIANCE MATRIX

	constant	time	timesq
constant	11.72906		
time	-5.86552	6.59372	
timesq	0.38927	-0.55909	0.05814

2.5 MULTILEVEL EXAMPLES

```
           LEVEL 2 CORRELATION MATRIX

               constant      time     timesq

constant        1.0000
time           -0.6670    1.0000
timesq          0.4714   -0.9030    1.0000

           LEVEL 1 COVARIANCE MATRIX

               constant

constant        3.07863

           LEVEL 1 CORRELATION MATRIX

               constant

constant        1.0000
```

- The fixed effects are all highly significant. There is an expected decrease of 0.2963 grams for every unit increase in the squared value of the time points. On the other hand, there is an estimated increase of 6.9056 grams with every increase of one week in time.

- The expected value of the weights of the mice at time point number 2 may thus be calculated as

$$\text{Expected WEIGHT}_{i2} = 4.1621 + 2.00(6.9056) + 4.00(0.2963)$$
$$= 16.7881 \text{ grams}.$$

- From the random part of the model it can be seen that all the estimates of the random coefficients at level 2 of the model are significant. This also holds for all the interaction terms at this level of the model. The correlation between TIME and TIMESQ is rather high, at -0.9030.

- Variation over measurements, on level 1 of the model, has been drastically reduced through the inclusion of the variable TIMESQ in the analysis. When comparing $-2\ln L$ for this model with that obtained for the linear growth model, a reduction of 473.7280 is noted, indicating that the inclusion of the variable TIMESQ significantly improved the fit of the model.

- The empirical Bayes residuals and their variances for the first five mice, as given in the file MOUSE4.BA2, is given in the table below.

1	1	5.9047	3.3066	CONSTANT
1	2	-3.6104	0.73621	TIME
1	3	0.36730	0.73356E-02	TIMESQ
2	1	0.16302	3.3066	CONSTANT
2	2	-1.1345	0.73621	TIME
2	3	0.13707	0.73356E-02	TIMESQ
3	1	-3.1320	3.3066	CONSTANT
3	2	2.3674	0.73621	TIME
3	3	-0.13993	0.73356E-02	TIMESQ
4	1	0.16759	3.3066	CONSTANT
4	2	-0.18088	0.73621	TIME
4	3	0.65171E-01	0.73356E-02	TIMESQ
5	1	-5.5456	3.3066	CONSTANT
5	2	0.96351	0.73621	TIME
5	3	0.85922E-01	0.73356E-02	TIMESQ

From this information the empirical Bayes estimates for any of the level-2 units may be computed. For the first five mice, these estimates are given in the following table.

Mouse no.	Fixed effect	Empirical Bayes estimate
1	CONSTANT	$4.1621 + 5.9047 = 10.0668$
	TIME	$6.9056 - 3.6104 = 3.2952$
	TIMESQ	$-0.2963 + 0.3673 = 0.0710$
2	CONSTANT	$4.1621 + 0.1630 = 4.3251$
	TIME	$6.9056 - 1.1345 = 5.7711$
	TIMESQ	$-0.2963 + 0.1371 = 0.1592$
3	CONSTANT	$4.1621 - 3.1320 = 1.0301$
	TIME	$6.9056 + 2.3674 = 9.2730$
	TIMESQ	$-0.2963 - 0.1399 = -0.4362$
4	CONSTANT	$4.1621 + 0.1676 = 4.3297$
	TIME	$6.9056 - 0.1809 = 6.7247$
	TIMESQ	$-0.2963 + 0.0652 = 0.2311$
5	CONSTANT	$4.1621 - 5.5456 = -1.3835$
	TIME	$6.9056 + 0.9635 = 7.8691$
	TIMESQ	$-0.2963 + 0.0859 = 0.2104$

2.5 MULTILEVEL EXAMPLES

The expected value of the weight of the first five mice at time point number 2 using the empirical Bayes estimates may thus be calculated as:

$$\text{WEIGHT}_{12} = 10.0668 + 2.00(3.2952) + 4.00(0.0710) = 16.9412 \text{ grams}$$

$$\text{WEIGHT}_{22} = 4.3251 + 2.00(5.7711) + 4.00(0.1592) = 15.2305 \text{ grams}$$

$$\text{WEIGHT}_{32} = 1.0301 + 2.00(9.2730) + 4.00(-0.4362) = 17.8313 \text{ grams}$$

$$\text{WEIGHT}_{42} = 4.3297 + 2.00(6.7247) + 4.00(0.2311) = 16.8547 \text{ grams}$$

$$\text{WEIGHT}_{52} = 9.7077 + 2.00(7.8961) + 4.00(-1.3835) = 19.9659 \text{ grams}$$

In the case of mice numbers 1, 3, 4, and 5, the estimated weights thus obtained are higher than previously calculated, while mouse number 2 is below the previously calculated value of 16.7881 units.

Finally, the residuals for the first 45 observations, that is the first five male mice, are considered. The following is an extract from the output file MOUSE4.RES:

```
 1    1    1    15.000    10.771     4.2286
 2    1    2    17.000    16.788     0.21183
 3    1    3    23.000    22.212     0.78768
 4    1    4    24.000    27.044    -3.0439
 5    1    5    26.000    31.283    -5.2829
 6    1    6    31.000    34.929    -3.9293
 7    1    7    37.000    37.983    -0.98314
 8    1    8    42.000    40.444     1.5556
 9    1    9    46.000    42.313     3.6869
10    2    1    11.000    10.771     0.22856
11    2    2    14.000    16.788    -2.7882
12    2    3    20.000    22.212    -2.2123
13    2    4    24.000    27.044    -3.0439
14    2    5    29.000    31.283    -2.2829
15    2    6    35.000    34.929     0.70698E-01
16    2    7    36.000    37.983    -1.9831
17    2    8    41.000    40.444     0.55561
18    2    9    43.000    42.313     0.68693
19    3    1    11.000    10.771     0.22856
20    3    2    16.000    16.788    -0.78817
21    3    3    24.000    22.212     1.7877
22    3    4    30.000    27.044     2.9561
23    3    5    39.000    31.283     7.7171
24    3    6    39.000    34.929     4.0707
25    3    7    48.000    37.983    10.017
26    3    8    47.000    40.444     6.5556
27    3    9    48.000    42.313     5.6869
28    4    1    13.000    10.771     2.2286
```

29	4	2	16.000	16.788	-0.78817
30	4	3	21.000	22.212	-1.2123
31	4	4	26.000	27.044	-1.0439
32	4	5	31.000	31.283	-0.28289
33	4	6	37.000	34.929	2.0707
34	4	7	44.000	37.983	6.0169
35	4	8	44.000	40.444	3.5556
36	4	9	44.000	42.313	1.6869
37	5	1	5.0000	10.771	-5.7714
38	5	2	13.000	16.788	-3.7882
39	5	3	20.000	22.212	-2.2123
40	5	4	26.000	27.044	-1.0439
41	5	5	32.000	31.283	0.71711
42	5	6	39.000	34.929	4.0707
43	5	7	45.000	37.983	7.0169
44	5	8	48.000	40.444	7.5556
45	5	9	52.000	42.313	9.6869

The largest residual for the male mice is for observation number 25 which represents the fifth measurement for mouse 3, where a residual of 10.017 is encountered. Plots of the residuals against the observation number are given in Figure 2.2. For a more detailed discussion of the analysis of residuals in a multilevel context, the user is referred to Goldstein (1987, pp. 21–26).

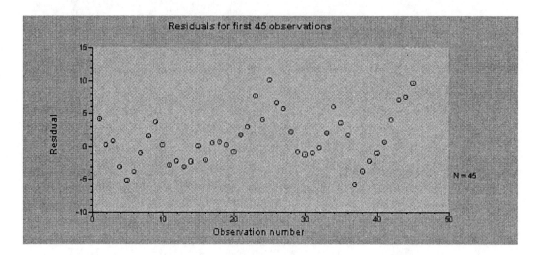

Figure 2.2   Plot of residuals for the first 45 records in the data set

2.5 MULTILEVEL EXAMPLES

Introducing a covariate while modeling non-linear growth

In this example we want to determine whether there is a significant difference between the growth pattern of the male and female mice, as modeled in the non-linear growth model discussed previously. This can be determined by adding the gender of the mice as covariate to the model fitted in the previous section.

The variable GENDER is introduced as covariate by modifying the FIXED command to the following:

```
FIXED = CONSTANT  TIME  TIMESQ  GENDER  GENDER*TIME  GENDER*TIMESQ;
```

Edit the previous file accordingly and save it under a different name: (MOUSE5.PR2). Run PRELIS with this input file to fit the model to the data. Convergence is reached after 4 iterations and the following output is obtained.

```
ITERATION NUMBER    4
```

	FIXED PART OF MODEL			
COEFFICIENTS	BETA-HAT	STD.ERR.	Z-VALUE	PR > |z|
constant	4.19133	0.44972	9.31996	0.00000
time	6.87771	0.29532	23.28885	0.00000
timesq	-0.29399	0.02900	-10.13902	0.00000
gender	-0.81492	0.44972	-1.81207	0.06998
gender *time	0.87235	0.29532	2.95389	0.00314
gender *timesq	-0.05947	0.02900	-2.05087	0.04028

| -2 LOG-LIKELIHOOD |

-2 LOG-LIKELIHOOD =   3389.53831478844

```
                    +----------------------+
                    | RANDOM PART OF MODEL |
                    +----------------------+

-----------------------------------------------------------------------
LEVEL 2                   TAU-HAT      STD.ERR.     Z-VALUE    PR > |Z|
-----------------------------------------------------------------------
constant/constant        11.07339       2.60146      4.25660    0.00002
time    /constant        -5.16047       1.46703     -3.51763    0.00044
time    /time             5.83687       1.11334      5.24265    0.00000
timesq  /constant         0.34138       0.13349      2.55735    0.01055
timesq  /time            -0.50761       0.10447     -4.85897    0.00000
timesq  /timesq           0.05464       0.01069      5.11098    0.00000

-----------------------------------------------------------------------
LEVEL 1                   TAU-HAT      STD.ERR.     Z-VALUE    PR > |Z|
-----------------------------------------------------------------------
constant/constant         3.07844       0.20464     15.04338    0.00000
```

LEVEL 2 COVARIANCE MATRIX

	constant	time	timesq
constant	11.07339		
time	-5.16047	5.83687	
timesq	0.34138	-0.50761	0.05464

LEVEL 2 CORRELATION MATRIX

	constant	time	timesq
constant	1.0000		
time	-0.6419	1.0000	
timesq	0.4389	-0.8989	1.0000

LEVEL 1 COVARIANCE MATRIX

	constant
constant	3.07844

LEVEL 1 CORRELATION MATRIX

	constant
constant	1.0000

- ❏ The coefficients of GENDER*TIME and GENDER*TIMESQ are significant at a 5 percent level, but the coefficient for GENDER*CONSTANT = GENDER is not.
- ❏ Only small changes are noticeable when the random part of the

2.5 MULTILEVEL EXAMPLES

model fitted is compared to the corresponding section of the output obtained in the third example.

❑ When comparing the values of $-2\ln L$ (likelihood function) for the two models, a reduction of 11.39 is noted.

❑ It can thus be concluded that, although the gender of the mice has no significance on the intercept denoted by the variable CONSTANT, there are significant differences between the growth patterns of male and female mice prior to physical maturity.

Complex variation at level 1 of the model

In the final example of a level-2 model, the linear growth model fitted previously is extended to include complex variation on both levels of the hierarchy. The term 'complex variation' refers to the existence of two or more random variables at the same level of the hierarchy. We include the variable TIME in this model to illustrate such a model.

We modify the RANDOM1 command previously used to

```
RANDOM1 = CONSTANT TIME;
```

This change in the level-1 covariance structure implies that the total variation at this level of the model can now be written as:

$$\text{Var}\left(\text{CONSTANT}_{ij} + \text{TIME}_{ij}\Phi_{(1)\text{TIME,TIME}}\right) =$$

$$\Phi_{(1)\text{CONSTANT,CONSTANT}} + 2\Phi_{(1)\text{TIME,CONSTANT}}(\text{TIME}_{ij}) + \Phi_{(1)\text{TIME,TIME}}(\text{TIME}_{ij})^2$$

The input file (MOUSE6.PR2) should contain the following:

```
OPTIONS OLS=YES CONVERGE=0.001000 MAXITER=25 OUTPUT=ALL ;
TITLE=Mouse data: variance decomposition;
SY=K:\LISREL83\MLEVELEX\MOUSE.PSF;
ID1=iden1;
ID2=iden2;
RESPONSE=weight;
FIXED=constant time;
RANDOM1=constant time;
RANDOM2=constant time;
```

The output obtained for this model is given below. This model needed 18 iterations for convergence. The MAXITER option of the OPTIONS command was used to increase the number of iterations from the default of 10 to 25.

```
ITERATION NUMBER 18
                        +----------------------+
                        | FIXED PART OF MODEL  |
                        +----------------------+

---------------------------------------------------------------------------
COEFFICIENTS            BETA-HAT     STD.ERR.     Z-VALUE      PR > |Z|
---------------------------------------------------------------------------
constant                 8.67635      0.46342     18.72259      0.00000
time                     4.30965      0.13067     32.98166      0.00000

                        +----------------------+
                        |   -2 LOG-LIKELIHOOD  |
                        +----------------------+

-2 LOG-LIKELIHOOD =     3822.83149832868

                        +----------------------+
                        | RANDOM PART OF MODEL |
                        +----------------------+

---------------------------------------------------------------------------
LEVEL 2                 TAU-HAT      STD.ERR.     Z-VALUE      PR > |Z|
---------------------------------------------------------------------------
constant/constant       13.35035      2.77510      4.81076      0.00000
time    /constant       -1.61832      0.61801     -2.61861      0.00883
time    /time            1.17651      0.21925      5.36601      0.00000

---------------------------------------------------------------------------
LEVEL 1                 TAU-HAT      STD.ERR.     Z-VALUE      PR > |Z|
---------------------------------------------------------------------------
constant/constant       13.73386      2.13539      6.43156      0.00000
time    /constant       -2.91233      0.54471     -5.34654      0.00000
time    /time            0.83668      0.12967      6.45252      0.00000

        LEVEL 2 COVARIANCE MATRIX

                constant      time

constant        13.35035
time            -1.61832     1.17651

        LEVEL 2 CORRELATION MATRIX

                constant      time

constant         1.0000
time            -0.4083      1.0000
```

```
                LEVEL 1 COVARIANCE MATRIX

                     constant        time

       constant      13.73386
       time          -2.91233       0.83668

                LEVEL 1 CORRELATION MATRIX

                     constant        time

       constant       1.0000
       time          -0.8591        1.0000
```

From the random part of the output it can be seen that there is an increase in the variance of the CONSTANT term on level 1 of the model when the coefficient for the variable TIME is also allowed to vary randomly over level 1 of the model. The error 1 variance, however, is also a function of the covariance between CONSTANT and TIME as shown on page 89. All coefficients are highly significant.

When the two values of $-2\ln L$ are compared for these models, a decrease of 50.43 is noted. The addition of the coefficient for the variable TIME on level 1 of the model thus seems to lead to an improved fit compared with the linear growth model. This implies that the level-1 error variances are heteroscedastic.

Conclusions

In the five examples discussed here, various models were considered for the analysis of repeated measurement data with a level-2 hierarchical structure. These models included a variance decomposition model, two linear growth models and a non-linear growth model. The inclusion of a covariate and the possibility of complex level-1 variation were also considered.

When the respective $-2\ln L$ values of these models are compared, the non-linear model with a covariate included had the lowest value, namely, 3389.5383. It would thus appear that the growth of the 82 mice up to physical maturity can best be described by a parabola with the gender of the mice as covariate. From Figure 2.1, however, it seems as if other non-linear functions for the modeling of the growth of the mice can also be considered.

2.5.2 Analysis of air traffic control data

The data used in this example are described by Kanfer & Ackerman (1989).[4] The data consists of information for 141 U.S. Air Force enlisted personnel. The personnel carried out a computerized air traffic controller task developed by Kanfer and Ackerman.

The subjects were instructed to accept planes into their hold pattern and land them safely and efficiently on one of four runways, varying in length and compass directions, according to rules governing plane movements and landing requirements. For each subject, the success of a series of between three and six 10-minute trials was recorded. The measurement employed was the number of correct landings per trial.

The Armed Services Vocational Battery (ASVB) was also administered to each subject. A global measure of cognitive ability, obtained from the sum of scores on 10 subscales, is included in the data.

The data for this example can be found in the KANFER.PSF file. The variable labels and first few records of this data file are shown below.

CONTROL	TIME	MEASURE	ABILITY	CONSTANT	TIMESQ
1.00	1.00	24.00	142.16	1.00	1.00
1.00	2.00	27.00	142.16	1.00	4.00
1.00	3.00	30.00	142.16	1.00	9.00
1.00	4.00	32.00	142.16	1.00	16.00
1.00	5.00	38.00	142.16	1.00	25.00
1.00	6.00	41.00	142.16	1.00	36.00
2.00	1.00	2.00	−7.63	1.00	1.00
2.00	2.00	3.00	−7.63	1.00	4.00
2.00	3.00	9.00	−7.63	1.00	9.00
2.00	4.00	13.00	−7.63	1.00	16.00
2.00	5.00	13.00	−7.63	1.00	25.00
2.00	6.00	14.00	−7.63	1.00	36.00
3.00	1.00	12.00	−67.43	1.00	1.00
3.00	2.00	18.00	−67.43	1.00	4.00
3.00	3.00	24.00	−67.43	1.00	9.00

---

[4]Permission for SSI to use the copyrighted raw data was provided by R. Kanfer and P.L. Ackerman. The data are from experiments reported in: Kanfer, R., & Ackerman, P.L. (1989).

The data remain the copyrighted property of Ruth Kanfer and Phillip L. Ackerman. Further publication or further dissemination of these data is not permitted without the expressed consent of the copyright owners.

2.5 MULTILEVEL EXAMPLES

The variables in the data set are:

CONTROL	The identifying number of the air traffic controller
TIME	The number of the trial (between 1 and 6)
MEASURE	The number of successful landings for the trial
ABILITY	The cognitive ability score (combined ASVB score)
CONSTANT	The intercept term, with value 1 throughout
TIMESQ	TIME*TIME, a quadratic term

Using this data, three models will be fitted:

- The first model, a variance decomposition model, will investigate the variation in the number of correct landings over subjects and also over measurements for each subject.
- In the next model, a non-linear growth model will be considered.
- Finally, the cognitive ability measure, a controller-related variable, will be introduced into the non-linear growth model.

Variance decomposition model for Air Force data

To start the input file, we select the defaults for output options, maximum number of iterations, and convergence criterion, then provide an optional title for the analysis.

The variable CONTROL, identifying the air traffic controller, is used as level-2 identification, as up to six measurements are available for each controller. The variable TIME, indicating the number of the trial, is used as level-1, or measurement, identification.

The number of successful landings per trial is represented by the variable MEASURE, which we select as the response variable for this particular model. We include the variable CONSTANT, representing the intercept term, as fixed effect in a similar way.

Finaly, the intercept term, as represented by the variable CONSTANT is selected as a random variable at both levels of the model.

The resulting input file (KANFER1.PR2) is as follows:

```
OPTIONS ;
TITLE=Kanfer and Ackerman data: Variance decomposition;
SY=K:\LISREL83\MLEVELEX\KANFER.PSF;
ID1=time;
ID2=control;
RESPONSE=measure;
FIXED=constant;
RANDOM1=constant;
RANDOM2=constant;
```

Partial output is given and discussed below.

```
            +---------------+
            | DATA SUMMARY  |
            +---------------+

NUMBER OF LEVEL 2 UNITS :    141
NUMBER OF LEVEL 1 UNITS :    840

N2 :      1      2      3      4      5      6      7      8
N1 :      6      6      6      6      6      6      6      6

N2 :      9     10     11     12     13     14     15     16
N1 :      6      6      6      6      6      6      6      6

N2 :     17     18     19     20     21     22     23     24
N1 :      6      6      6      6      6      6      6      6

N2 :     25     26     27     28     29     30     31     32
N1 :      6      6      6      6      6      6      6      6

N2 :     33     34     35     36     37     38     39     40
N1 :      6      6      6      6      6      6      6      6

N2 :     41     42     43     44     45     46     47     48
N1 :      6      6      6      6      6      6      6      6

N2 :     49     50     51     52     53     54     55     56
N1 :      6      6      6      6      6      6      6      6

N2 :     57     58     59     60     61     62     63     64
N1 :      6      6      6      6      6      6      6      6

N2 :     65     66     67     68     69     70     71     72
N1 :      6      6      6      6      6      6      6      6

N2 :     73     74     75     76     77     78     79     80
N1 :      6      6      6      6      6      6      6      6

N2 :     81     82     83     84     85     86     87     88
N1 :      6      6      6      6      6      6      6      6
```

2.5 MULTILEVEL EXAMPLES

```
N2 :    89    90    91    92    93    94    95    96
N1 :     6     6     6     6     6     6     6     6

N2 :    97    98    99   100   101   102   103   104
N1 :     6     6     6     6     6     6     6     6

N2 :   105   106   107   108   109   110   111   112
N1 :     6     6     6     6     6     6     6     6

N2 :   113   114   115   116   117   118   119   120
N1 :     6     6     6     6     6     6     6     6

N2 :   121   122   123   124   125   126   127   128
N1 :     6     6     6     6     6     6     6     6

N2 :   129   130   131   132   133   134   135   136
N1 :     6     6     6     6     6     6     6     6

N2 :   137   138   139   140   141
N1 :     6     5     5     5     3
```

ITERATION NUMBER   4

+----------------------+
| FIXED PART OF MODEL |
+----------------------+

COEFFICIENTS	BETA-HAT	STD.ERR.	Z-VALUE	PR > |Z|
constant	26.15108	0.66707	39.20293	0.00000

+----------------------+
| -2 LOG-LIKELIHOOD |
+----------------------+

-2 LOG-LIKELIHOOD =   6380.91436790376

+----------------------+
| RANDOM PART OF MODEL |
+----------------------+

LEVEL 2	TAU-HAT	STD.ERR.	Z-VALUE	PR > |Z|
constant/constant	47.22948	7.51714	6.28290	0.00000

LEVEL 1	TAU-HAT	STD.ERR.	Z-VALUE	PR > |Z|
constant/constant	92.17651	4.93041	18.69552	0.00000

```
            LEVEL 2 COVARIANCE MATRIX

                    constant

    constant        47.22948

            LEVEL 2 CORRELATION MATRIX

                    constant

    constant        1.0000

            LEVEL 1 COVARIANCE MATRIX

                    constant

    constant        92.17651

            LEVEL 1 CORRELATION MATRIX

                    constant

    constant        1.0000
```

- From the *Data Summary*, we see that we have complete data for all but the last four air traffic controllers.
- The mean number of successful trials, as obtained from the output for the fixed part of the model, is 26.15108. From exploratory analysis of this data done prior to analysis, it is known that the number of successful trials ranged between 0 and 57.
- From the output for the random part of the model, we see that the variation over measurements is approximately twice the variation over the controllers (47.2298 versus 92.1765). Most of the variation is within controllers (over measurements), but it is noted that variation in intercepts was significant at both levels.
- A value of 6380.914 was obtained for $-2\ln L$.

In the next model, we will consider a non-linear growth model and compare results with the results obtained here.

Non-linear model for air traffic data

In the example above, a basic model for the analysis of the air traffic data of Kanfer & Ackerman was considered.

2.5 MULTILEVEL EXAMPLES

In data of this nature, it can be expected that the number of successful landings per trial may be influenced by previous experience. In order to take this into account, we now introduce the order of the measurements into the model. The variable TIME indicates the number of the trial, and TIMESQ the quadratic term equal to TIME*TIME.

To fit such a model, the FIXED and RANDOM2 commands used previously have to be changed to:

```
FIXED = constant time timesq;
RANDOM2 = constant time timesq;
```

These changes can be made by directly editing the KANFER1.PR2 file used in the previous analysis, and saving it as KANFER2.PR2:

```
OPTIONS ;
TITLE=Kanfer and Ackerman data: Non-linear model;
SY=K:\LISREL83\MLEVELEX\KANFER.PSF;
ID1=time;
ID2=control;
RESPONSE=measure;
FIXED=constant time timesq;
RANDOM1=constant;
RANDOM2=constant time timesq;
```

Partial output for this model follows.

```
ITERATION NUMBER     4
```

```
+-----------------------+
| FIXED PART OF MODEL   |
+-----------------------+
```

COEFFICIENTS	BETA-HAT	STD.ERR.	Z-VALUE	PR > |Z|
constant	1.67806	0.87343	1.92123	0.05470
time	11.48966	0.50397	22.79838	0.00000
timesq	-1.03583	0.06141	-16.86761	0.00000

```
+-----------------------+
|  -2 LOG-LIKELIHOOD    |
+-----------------------+
```

-2 LOG-LIKELIHOOD =   5095.46585430938

```
+----------------------+
| RANDOM PART OF MODEL |
+----------------------+
```

LEVEL 2	TAU-HAT	STD.ERR.	Z-VALUE	PR > |Z|
constant/constant	77.02126	12.98193	5.93296	0.00000
time /constant	-24.73085	6.50474	-3.80197	0.00014
time /time	22.66462	4.35831	5.20033	0.00000
timesq /constant	2.26771	0.76607	2.96017	0.00307
timesq /time	-2.38073	0.52370	-4.54595	0.00001
timesq /timesq	0.27282	0.06566	4.15473	0.00003

LEVEL 1	TAU-HAT	STD.ERR.	Z-VALUE	PR > |Z|
constant/constant	9.49299	0.65567	14.47823	0.00000

LEVEL 2 COVARIANCE MATRIX

	constant	time	timesq
constant	77.02126		
time	-24.73085	22.66462	
timesq	2.26771	-2.38073	0.27282

LEVEL 2 CORRELATION MATRIX

	constant	time	timesq
constant	1.0000		
time	-0.5919	1.0000	
timesq	0.4947	-0.9574	1.0000

LEVEL 1 COVARIANCE MATRIX

	constant
constant	9.49299

LEVEL 1 CORRELATION MATRIX

	constant
constant	1.0000

CONVERGENCE REACHED IN 4 ITERATIONS

From the output above, we see that:

- Both the TIME and TIMESQ variables, introduced as fixed effects in this model, have coeffcents which are highly significant. From the large coefficient for TIME (11.48966) we see that an increase of 11 successful landings is expected between successive trials.
- From the random part of the output, it follows that all the estimates of the random coefficients on level 2 of the model are significant. This is also true for all the interaction terms on this level of the model.
- The estimate of the coefficient of the level-1 variance is now 9.49299, compared to the coefficient for the similar term in the previous model of 92.17651, where TIME and TIMESQ were not included in the model. Variation over measurements has been drastically reduced in the extended model.
- For this model, a $-2\ln L$ value of 5095.466 was obtained, compared to the previous model where $-2\ln L$ was 6380.914. In the previous model, 3 parameters were estimated, while for the extended model 10 parameters were estimated. The inclusion of TIME and TIMESQ significantly improved the fit of the model.

In the two models considered thus far, the measure of cognitive ability available for each controller has not been considered. In the third and final analysis of this data, we will include this measure in the model considered in this section.

Including additional variables in the air traffic data analysis

In the previous example a non-linear model was considered.

Recall that the data set also includes a measure of cognitive ability (composite ASVB score) for each of the 141 air traffic controllers, denoted by ABILITY in the file KANFER.PSF. We now include this variable as a fixed effect in the model.

To add ABILITY to the model, the input file used in the previous example may be edited. After it has been added to the fixed variables field, the input file for this analysis is as follows (KANFER3.PR2).

```
OPTIONS ;
TITLE=Kanfer and Ackerman data: Non-linear model with covariate;
SY=K:\LISREL83\MLEVELEX\KANFER.PSF;
ID1=time;
ID2=control;
RESPONSE=measure;
FIXED=constant time timesq ability;
RANDOM1=constant;
RANDOM2=constant time timesq;
```

Partial output for this analysis is shown below.

ITERATION NUMBER   4

+------------------------+
| FIXED PART OF MODEL |
+------------------------+

COEFFICIENTS	BETA-HAT	STD.ERR.	Z-VALUE	PR > |Z|
constant	1.67596	0.78914	2.12379	0.03369
time	11.49203	0.50394	22.80458	0.00000
timesq	-1.03639	0.06140	-16.87917	0.00000
ability	0.03703	0.00519	7.12990	0.00000

+------------------------+
| -2 LOG-LIKELIHOOD |
+------------------------+

-2 LOG-LIKELIHOOD =   5052.79195898966

+------------------------+
| RANDOM PART OF MODEL |
+------------------------+

LEVEL 2	TAU-HAT	STD.ERR.	Z-VALUE	PR > |Z|
constant/constant	57.26989	10.66634	5.36921	0.00000
time /constant	-22.58400	5.99981	-3.76412	0.00017
time /time	22.66330	4.35771	5.20074	0.00000
timesq /constant	2.09446	0.70778	2.95921	0.00308
timesq /time	-2.38017	0.52358	-4.54592	0.00001
timesq /timesq	0.27273	0.06565	4.15457	0.00003

2.5 MULTILEVEL EXAMPLES

```
--------------------------------------------------------------
  LEVEL 1                    TAU-HAT    STD.ERR.   Z-VALUE   PR > |Z|
--------------------------------------------------------------
  constant/constant          9.49057    0.65551   14.47819   0.00000
```

```
            LEVEL 2 COVARIANCE MATRIX

              constant        time       timesq

constant      57.26989
time         -22.58400      22.66330
timesq         2.09446      -2.38017     0.27273


            LEVEL 2 CORRELATION MATRIX

              constant        time       timesq

constant       1.0000
time          -0.6269       1.0000
timesq         0.5300      -0.9574       1.0000


            LEVEL 1 COVARIANCE MATRIX

              constant

constant      9.49057


            LEVEL 1 CORRELATION MATRIX

              constant

constant       1.0000
```

CONVERGENCE REACHED IN 4 ITERATIONS

From the output above, we see that:

- The effect of ability on the number of successful landings is positive, indicating significant increases in the expected number of successful landings per trial with increase in ability values given that the time components are held constant. The largest effect is that of the order of trials, indicating that a higher number of succesful landings are expected in a following trial.

- From the large coefficient for TIME (11.4920), we see that there is a positive increase of 11 successful landings expected between successive trials.
- From the random part of the output, it follows that all the estimates of the random coefficients on level 2 of the model are significant. This is also true for all the interaction terms at this level of the model. Note that the coefficient for the intercept at this level has been reduced to 57.26989, compared to 77.02126 in the previous model, where the cognitive ability measure was not included. The introduction of this measure into the model leads to a reduction in the variation of mean outcome scores over controllers.
- The estimate of the coefficient of the level-1 variance is now 9.4057, which is very similar to the estimate obtained for the model without ABILITY. Inclusion of this variable did not contribute to the reduction in the variation over measurements nested within each controller.
- For this model, a $-2\ln L$ value of 5052.792 was obtained, compared to the previous model where $-2\ln L$ was 5095.466. As only one extra parameter was estimated in this model, we conclude that the inclusion of ABILITY in the non-linear model significantly improved the fit of the model.

2.5.3 A multivariate analysis of educational data

The data set used in this section forms part of the data library of the Multilevel Project at the University of London and comes from the Junior School Project (Mortimer, *et al.*, 1988).

Mathematics and language tests were administered in three consecutive years to more than 1000 students from 50 primary schools, which were randomly selected from primary schools maintained by the Inner London Education Authority.

The following variables are available in the data file JSP.PSF:

The school number (SCHOOL) is used as the level-3 identification variable, with Student ID (STUDENT) as the level-2 identification variable. The level-1 units are the scores obtained by a student for the mathematics and language tests, as represented by MATH1 to MATH3 and ENG1 to ENG3.

2.5 MULTILEVEL EXAMPLES

SCHOOL	School code (1 to 50)
CLASS	Class code (1 to 4)
STUDENT	Student ID (1 to 1402)
GENDER	Gender (boy=1; girl=0)
RAVENS	Ravens test score in year 1 (score 4–36)
MATH1	Score on mathematics test in year 1 (score 1–40)
MATH2	Score on mathematics test in year 2 (score 1–40)
MATH3	Score on mathematics test in year 3 (score 1–40)
ENG1	Score on language test in year 1 (score 0–98)
ENG2	Score on language test in year 2 (score 0–98)
ENG3	Score on language test in year 3 (score 0–98)
CONSTANT	Intercept, value=1 throughout

The aim of this analysis is to examine the variation in test scores over students. It would also be interesting to determine the extent to which schools vary with respect to the response variable(s). One of the main benefits of analyzing different responses simultaneously in one multivariate analysis is that the way in which measurements relate to the explanatory variables can be directly explored. The gender, Ravens test score, and an intercept term (GENDER, RAVENS, and CONSTANT, respectively) will be used as explanatory variables in the second of the models described in this section.

As the residual covariance matrices for the second and third levels of the hierarchy are also obtained from this analysis, differences between coefficients of explanatory variables for different responses can be studied.

Finally, each respondent does not have to be measured on each response, as is the case for the data set we consider here, where scores for all three years are not available for all students. This type of model is one in which the MISSING_DEP and MISSING_DAT commands can be used to good advantage, as will be shown. In the case of missing responses, a multilevel multivariate analysis of the responses that are available for respondents can be used to provide information in the estimation of those that are missing.

The first fifteen observations in the JSP.PSF file are shown on the next page.

SCHOOL	STUDENT	GENDER	RAVENS	MATH1	MATH2	MATH3	ENG1	ENG2	ENG3	CONSTANT
1.00	1.00	0.00	23.00	23.00	24.00	23.00	72.00	80.00	39.00	1.00
1.00	2.00	1.00	15.00	14.00	11.00	-9.00	7.00	17.00	-9.00	1.00
1.00	3.00	1.00	22.00	36.00	32.00	39.00	88.00	89.00	83.00	1.00
1.00	4.00	1.00	14.00	24.00	26.00	32.00	12.00	25.00	12.00	1.00
1.00	5.00	0.00	19.00	22.00	23.00	-9.00	67.00	78.00	-9.00	1.00
1.00	6.00	0.00	16.00	19.00	23.00	11.00	52.00	76.00	19.00	1.00
1.00	7.00	1.00	17.00	22.00	22.00	26.00	37.00	68.00	31.00	1.00
1.00	8.00	0.00	21.00	18.00	29.00	28.00	57.00	86.00	40.00	1.00
1.00	9.00	1.00	30.00	30.00	31.00	-9.00	42.00	59.00	-9.00	1.00
1.00	10.00	0.00	25.00	29.00	29.00	-9.00	46.00	79.00	-9.00	1.00
1.00	11.00	0.00	32.00	31.00	28.00	32.00	69.00	84.00	50.00	1.00
1.00	12.00	0.00	15.00	18.00	26.00	-9.00	54.00	74.00	-9.00	1.00
1.00	13.00	0.00	25.00	23.00	-9.00	27.00	63.00	-9.00	39.00	1.00
1.00	14.00	0.00	29.00	39.00	35.00	36.00	83.00	88.00	80.00	1.00
1.00	15.00	0.00	34.00	24.00	30.00	33.00	37.00	44.00	37.00	1.00

One line of information is given for each student. Note that, for the second and fourth students, no mathematics score was available in the third year of the study (MATH3). A missing data code of −9 was assigned to all missing values on both explanatory and response variables in the data set.

In order to perform a multilevel analysis, we need one line of information for each level-1 unit, in this case each of the six test scores. The data manipulation required in creating this revised data file format will be performed automatically by the program in the case of a multivariate model. Six dummy variables are created for each of the explanatory variables used. For the explanatory variable GENDER, for example, the dummy variables GENDER1, GENDER2, ..., GENDER6 will denote the gender effect for each of the response variables.

Two models will be fitted and discussed:

- A variance decomposition model
- Adding explanatory variables to the model

A variance decomposition model

As a first step, we provide a title for our analysis. No specifications (other than the defaults) are needed for the OPTIONS command, and thus the default convergence criterion and maximum number of iterations will be used. Only the default output file will be produced.

2.5 MULTILEVEL EXAMPLES

In the case of a multivariate model, only levels 2 and 3 of the hierarchy have to be identified, because the actual measurements for each level-2 unit serve as level-1 units. In this case, we select SCHOOL as the level-3 identification variable. The student ID (STUDENT) is selected as the level-2 identification variable.

We choose the six variables MATH1 to MATH3 and ENG1 to ENG3 as response variables.

In this model, we wish to start our analysis of the data with a look at the differences in intercepts over students and schools. Note that in this analysis, no RANDOM commands should be included. These commands are only required when running a multivariate model when more variables than just the intercept is required at higher levels of the hierarchy. If no RANDOM commands are included, the inclusion of an intercept term on higher levels is automatically assumed by the program and dummy variables for the explanatory predictors given in the FIXED command are generated.

The resulting input file (JSP1.PR2) is given below.

Note, that the optional MISSING_DEP command is used to identify -9 as the missing data code for all response variables.

```
OPTIONS ;
TITLE=Multivariate Analysis of Education Data;
SY=K:\LISREL83\MLEVELEX\JSP.PSF;
ID2=student;
ID3=school;
RESPONSE=math1 math2 math3 eng1 eng2 eng3;
FIXED=constant;
MISSING_DEP=-9;
```

The output for this model is written to the default output file JSP1.OUT. Partial output is given below.

```
                          +--------------+
                          | DATA SUMMARY |
                          +--------------+

NUMBER OF LEVEL 3 UNITS :      49
NUMBER OF LEVEL 2 UNITS :    1192
NUMBER OF LEVEL 1 UNITS :    6472
```

```
ID3 :     1      2      3      4      5      6      7      8
N2  :    34     13     21     24     29     24     15     31
N1  :   184     72     96    144    166    120     78    174

ID3 :     9     10     11     12     13     14     15     16
N2  :    22     14     12     28     25     15     24     19
N1  :   130     48     70    154    138     86    106    106

ID3 :    17     18     19     20     21     22     23     24
N2  :     7     20     17     15     32     16     20     21
N1  :    40    110     88     84    184     92    116    126

ID3 :    25     26     27     28     29     30     31     32
N2  :    18     29     27     18     25     37     36     26
N1  :    94    156    146     96    124    182    214    148

ID3 :    33     34     35     36     37     38     39     40
N2  :    46     34     19     32     22     14     16     13
N1  :   262    184    106    178    118     78     90     68

ID3 :    41     42     43     44     45     46     47     48
N2  :    13     47      6     17     19     37     72     25
N1  :    70    262     32     88    102    204    412    144

ID3 :    49
N2  :    46
N1  :   202
```

[Output omitted]

Multivariate Analysis of Education Data

ITERATION NUMBER   8

```
                    +------------------------+
                    |  FIXED PART OF MODEL   |
                    +------------------------+

-------------------------------------------------------------------------------
COEFFICIENTS          BETA-HAT       STD.ERR.       Z-VALUE       PR > |Z|
-------------------------------------------------------------------------------
constan1              24.90370       0.33546       74.23792       0.00000
constan2              24.87234       0.40108       62.01311       0.00000
constan3              30.04909       0.37761       79.57736       0.00000
constan4              47.15338       1.32158       35.67945       0.00000
constan5              64.96594       1.22017       53.24352       0.00000
constan6              40.71988       1.38296       29.44399       0.00000

                    +------------------------+
                    |   -2 LOG-LIKELIHOOD    |
                    +------------------------+

-2 LOG-LIKELIHOOD =      45991.2578958991
```

2.5 MULTILEVEL EXAMPLES

```
+----------------------+
| RANDOM PART OF MODEL |
+----------------------+
```

LEVEL 3		TAU-HAT	STD.ERR.	Z-VALUE	PR > |z|
math1	/math1	3.31028	1.10053	3.00791	0.00263
math2	/math1	2.29203	1.09364	2.09579	0.03610
math2	/math2	5.23047	1.57789	3.31486	0.00092
math3	/math1	2.36877	1.02957	2.30075	0.02141
math3	/math2	3.13485	1.25421	2.49945	0.01244
math3	/math3	4.79640	1.39721	3.43285	0.00060
eng1	/math1	9.95162	3.74223	2.65927	0.00783
eng1	/math2	9.49602	4.22124	2.24958	0.02448
eng1	/math3	9.95900	4.00265	2.48810	0.01284
eng1	/eng1	59.32680	17.15964	3.45735	0.00055
eng2	/math1	9.93290	3.47650	2.85716	0.00427
eng2	/math2	11.41833	4.08446	2.79555	0.00518
eng2	/math3	11.59906	3.87949	2.98984	0.00279
eng2	/eng1	42.49228	14.15024	3.00294	0.00267
eng2	/eng2	53.08595	14.64628	3.62453	0.00029
eng3	/math1	10.16830	3.82941	2.65532	0.00792
eng3	/math2	10.71620	4.43202	2.41790	0.01561
eng3	/math3	13.71315	4.43882	3.08937	0.00201
eng3	/eng1	45.00410	15.55552	2.89313	0.00381
eng3	/eng2	51.61419	15.11775	3.41415	0.00064
eng3	/eng3	71.08796	18.81475	3.77831	0.00016

LEVEL 2		TAU-HAT	STD.ERR.	Z-VALUE	PR > |z|
math1	/math1	47.17671	1.99173	23.68631	0.00000
math2	/math1	38.63798	1.91420	20.18492	0.00000
math2	/math2	55.45831	2.35660	23.53314	0.00000
math3	/math1	31.21276	1.65633	18.84448	0.00000
math3	/math2	36.54802	1.84653	19.79278	0.00000
math3	/math3	41.33874	1.86018	22.22293	0.00000
eng1	/math1	109.03175	5.79022	18.83032	0.00000
eng1	/math2	109.44780	6.15532	17.78100	0.00000
eng1	/math3	88.00030	5.36198	16.41191	0.00000
eng1	/eng1	549.41232	23.15761	23.72492	0.00000
eng2	/math1	88.73168	4.92512	18.01613	0.00000
eng2	/math2	94.99391	5.32980	17.82315	0.00000
eng2	/math3	79.52971	4.69229	16.94901	0.00000
eng2	/eng1	388.36174	18.26625	21.26117	0.00000
eng2	/eng2	409.27297	17.36258	23.57213	0.00000
eng3	/math1	94.37662	5.19417	18.16973	0.00000
eng3	/math2	99.80683	5.60600	17.80356	0.00000
eng3	/math3	86.38268	4.95170	17.44505	0.00000
eng3	/eng1	382.76743	18.73683	20.42861	0.00000
eng3	/eng2	317.45309	15.98496	19.85949	0.00000
eng3	/eng3	425.49404	18.97401	22.42510	0.00000

```
LEVEL 3 COVARIANCE MATRIX

            math1      math2      math3      eng1       eng2       eng3

math1       3.31028
math2       2.29203    5.23047
math3       2.36877    3.13485    4.79640
eng1        9.95162    9.49602    9.95900   59.32680
eng2        9.93290   11.41833   11.59906   42.49228   53.08595
eng3       10.16830   10.71620   13.71315   45.00410   51.61419   71.08796

LEVEL 3 CORRELATION MATRIX

            math1      math2      math3      eng1       eng2       eng3

math1       1.0000
math2       0.5508     1.0000
math3       0.5945     0.6259     1.0000
eng1        0.7101     0.5391     0.5904     1.0000
eng2        0.7493     0.6852     0.7269     0.7572     1.0000
eng3        0.6629     0.5557     0.7426     0.6930     0.8402     1.0000

LEVEL 2 COVARIANCE MATRIX

            math1      math2      math3      eng1       eng2       eng3

math1       47.17671
math2       38.63798   55.45831
math3       31.21276   36.54802   41.33874
eng1       109.03175  109.44780   88.00030  549.41232
eng2        88.73168   94.99391   79.52971  388.36174  409.27297
eng3        94.37662   99.80683   86.38268  382.76743  317.45309  425.49404

LEVEL 2 CORRELATION MATRIX

            math1      math2      math3      eng1       eng2       eng3

math1       1.0000
math2       0.7554     1.0000
math3       0.7068     0.7633     1.0000
eng1        0.6772     0.6270     0.5839     1.0000
eng2        0.6386     0.6305     0.6114     0.8190     1.0000
eng3        0.6661     0.6497     0.6513     0.7917     0.7607     1.0000

CONVERGENCE REACHED IN  8 ITERATIONS
```

❑ From the *Data Summary*, we see that data from 1192 students from 49 schools were used in the analysis. The number of level-1 units (*i.e.*, measurements) per school ranged from 32 in the case of school number 43 to 412 for school number 47.

❑ From the output for the fixed part of the model, it can be seen that

2.5 MULTILEVEL EXAMPLES

all six fixed effects are highly significant, indicating significant differences in the six measurements over the students.

- There is significant variation in the mean effects of the six response variables over both schools and students. The variation over students (level 2) is higher than over schools. At the student level, the largest variation is in the mean effects for the language tests, ranging between 59.326 (for measurements from the first year), to 71.088 (for measurements from the third year of the study). The same tendency is observed for the mean effects of language test scores over schools, with the highest variation recorded for the language test score from the third year of the study. Keep in mind that the range of math scores is 40 while the range for language scores is 98.

- The $-2\ln L$ value for this analysis at convergence after 7 iterations was 45991.2579.

Adding explanatory variables to the model

Using the model discussed in the previous section as point of departure, we now proceed to add fixed effects to the model. The variables GENDER and RAVENS, indicating the gender of a student and the student's score on the Ravens test in the first year of the study, respectively, are added to the FIXED command.

This is how the FIXED command should look:

```
FIXED = CONSTANT GENDER RAVENS;
```

To make this change, the input file previously used (JSP1.PR2) can be edited directly, and then saved as the new input file JSP2.PR2:

```
OPTIONS ;
TITLE=Multivariate Analysis of Education Data;
SY=K:\LISREL83\MLEVELEX\JSP.PSF;
ID2=student;
ID3=school;
RESPONSE=math1 math2 math3 eng1 eng2 eng3;
FIXED=constant gender ravens;
MISSING_DEP=-9;
```

For this model the following output was obtained.

```
ITERATION NUMBER     7
```

+----------------------+
| FIXED PART OF MODEL |
+----------------------+

COEFFICIENTS	BETA-HAT	STD.ERR.	Z-VALUE	PR > |Z|
constan1	7.68926	0.80046	9.60607	0.00000
constan2	6.48252	0.90269	7.18136	0.00000
constan3	14.84541	0.84586	17.55061	0.00000
constan4	2.64745	2.90410	0.91163	0.36197
constan5	28.95062	2.57309	11.25128	0.00000
constan6	-2.58204	2.74794	-0.93963	0.34741
gender1	-0.48513	0.33815	-1.43466	0.15138
gender2	-0.79986	0.37167	-2.15205	0.03139
gender3	-0.45790	0.34277	-1.33588	0.18159
gender4	-10.57862	1.20875	-8.75172	0.00000
gender5	-9.21809	1.06819	-8.62960	0.00000
gender6	-7.02657	1.10494	-6.35926	0.00000
ravens1	0.69639	0.02943	23.66040	0.00000
ravens2	0.74945	0.03248	23.07353	0.00000
ravens3	0.61505	0.03029	20.30821	0.00000
ravens4	1.97944	0.10546	18.77035	0.00000
ravens5	1.61390	0.09328	17.30225	0.00000
ravens6	1.86293	0.09754	19.09957	0.00000

+----------------------+
| -2 LOG-LIKELIHOOD |
+----------------------+

-2 LOG-LIKELIHOOD =   45321.6134304450

+----------------------+
| RANDOM PART OF MODEL |
+----------------------+

LEVEL 3		TAU-HAT	STD.ERR.	Z-VALUE	PR > |Z|
math1	/math1	2.35970	0.77231	3.05538	0.00225
math2	/math1	1.47571	0.79018	1.86756	0.06182
math2	/math2	4.59531	1.29317	3.55353	0.00038
math3	/math1	1.45284	0.73222	1.98414	0.04724
math3	/math2	2.34726	0.97705	2.40240	0.01629
math3	/math3	3.94231	1.11612	3.53215	0.00041
eng1	/math1	5.89266	2.50131	2.35583	0.01848
eng1	/math2	5.10887	3.00832	1.69825	0.08946

2.5 MULTILEVEL EXAMPLES

eng1	/math3	5.75844	2.83649	2.03013	0.04234
eng1	/eng1	40.46361	11.99950	3.37211	0.00075
eng2	/math1	5.42374	2.23795	2.42353	0.01537
eng2	/math2	6.60379	2.82875	2.33453	0.01957
eng2	/math3	7.26912	2.69805	2.69422	0.00706
eng2	/eng1	22.52145	9.10418	2.47375	0.01337
eng2	/eng2	33.03809	9.66632	3.41786	0.00063
eng3	/math1	5.14461	2.49143	2.06492	0.03893
eng3	/math2	5.67697	3.12932	1.81413	0.06966
eng3	/math3	9.01133	3.15762	2.85384	0.00432
eng3	/eng1	22.93497	10.12809	2.26449	0.02354
eng3	/eng2	30.00480	9.77962	3.06810	0.00215
eng3	/eng3	47.71171	12.97810	3.67632	0.00024

LEVEL 2		TAU-HAT	STD.ERR.	Z-VALUE	PR > |Z|
math1	/math1	31.84368	1.34833	23.61706	0.00000
math2	/math1	22.14572	1.23554	17.92388	0.00000
math2	/math2	37.70830	1.61030	23.41693	0.00000
math3	/math1	17.66778	1.10211	16.03093	0.00000
math3	/math2	21.92750	1.24318	17.63828	0.00000
math3	/math3	29.34575	1.34391	21.83617	0.00000
eng1	/math1	65.59655	3.92248	16.72322	0.00000
eng1	/math2	62.26680	4.16530	14.94895	0.00000
eng1	/math3	49.65248	3.75896	13.20909	0.00000
eng1	/eng1	406.75188	17.17856	23.67788	0.00000
eng2	/math1	53.71682	3.40837	15.76025	0.00000
eng2	/math2	56.85501	3.69230	15.39826	0.00000
eng2	/math3	48.64973	3.37484	14.41540	0.00000
eng2	/eng1	272.02299	13.42728	20.25897	0.00000
eng2	/eng2	314.33951	13.37559	23.50099	0.00000
eng3	/math1	53.74137	3.50517	15.33203	0.00000
eng3	/math2	55.52744	3.78879	14.65570	0.00000
eng3	/math3	50.44039	3.44613	14.63681	0.00000
eng3	/eng1	256.17082	13.47180	19.01534	0.00000
eng3	/eng2	214.35205	11.70207	18.31744	0.00000
eng3	/eng3	310.90414	14.08313	22.07635	0.00000

LEVEL 3 COVARIANCE MATRIX

	math1	math2	math3	eng1	eng2	eng3
math1	2.35970					
math2	1.47571	4.59531				
math3	1.45284	2.34726	3.94231			
eng1	5.89266	5.10887	5.75844	40.46361		
eng2	5.42374	6.60379	7.26912	22.52145	33.03809	
eng3	5.14461	5.67697	9.01133	22.93497	30.00480	47.71171

```
LEVEL 3 CORRELATION MATRIX

          math1     math2     math3     eng1      eng2      eng3
math1    1.0000
math2    0.4481    1.0000
math3    0.4763    0.5515    1.0000
eng1     0.6030    0.3747    0.4559    1.0000
eng2     0.6143    0.5360    0.6369    0.6160    1.0000
eng3     0.4849    0.3834    0.6571    0.5220    0.7557    1.0000

LEVEL 2 COVARIANCE MATRIX

          math1     math2     math3     eng1       eng2       eng3
math1    31.84368
math2    22.14572  37.70830
math3    17.66778  21.92750  29.34575
eng1     65.59655  62.26680  49.65248 406.75188
eng2     53.71682  56.85501  48.64973 272.02299  314.33951
eng3     53.74137  55.52744  50.44039 256.17082  214.35205  310.90414

LEVEL 2 CORRELATION MATRIX

          math1     math2     math3     eng1      eng2      eng3
math1    1.0000
math2    0.6391    1.0000
math3    0.5780    0.6592    1.0000
eng1     0.5764    0.5028    0.4545    1.0000
eng2     0.5369    0.5222    0.5065    0.7607    1.0000
eng3     0.5401    0.5128    0.5281    0.7204    0.6857    1.0000

CONVERGENCE REACHED IN   7 ITERATIONS
```

Results for the fixed part of the model show that:

- The expected score of girls (GENDER=0) for all test scores is higher than the expected score for boys (GENDER=1). For boys (GENDER=1), the coefficients GENDER4 to GENDER6 are negative and highly significant. In the case of the mathematics test scores (MATH1 to MATH3 as represented by GENDER1 to GENDER3) the effects are smaller. Keep in mind that the range of scores differed between the mathematics and language tests.
- The effects of the RAVENS test are positive and highly significant for all six response variables, with the largest effects for the mathematics tests. An increase of one unit in the RAVENS test score implies

an expected increase in the third year language test of 0.61 and an expected increase of 1.86 for the expected third year mathematics test score.

Results for the random part of the model are consistent with the results for the previous model fitted, with larger variation for the three language tests and, in general, more variation over students than over schools.

The $-2\ln L$ recorded for this model is 45321.61343. When compared to the $-2\ln L$ of 45991.2579 obtained previously, a marked decrease is observed. In the first model, 48 parameters were estimated, compared to 60 parameters for the model discussed here. The introduction of the GENDER and RAVENS variables have contributed significantly to the explanation of variance in the response variables.

2.5.4 Analysis of CPC survey data

In this section, data from the March 1995 Current Population Survey are used. The data set is a subset of data obtained from the Data Library at the Department of Statistics at UCLA. A small number of demographic variables for two occupation groups was extracted, and all analyses are based on unweighted data.

Only respondents between the ages of twenty-one and sixty-five, who held full time positions in 1994 and had an annual income of US$1 or more were considered. The two groups we will focus on here are defined as follows:

Educational sector	Respondents with professional specialty in the educational sector
Construction sector	Operators, fabricators, and laborers in the construction sector

The variable GROUP in the PRELIS system file INCOME.PSF represents the groups, with GROUP=0 for the respondents in the construction sector and GROUP=1 for respondents in the educational sector. A 3–D bar chart showing the sample sizes of the two groups,[5] is shown in Figure 2.3.

---

[5]As obtained with the Graphs option in the Windows version

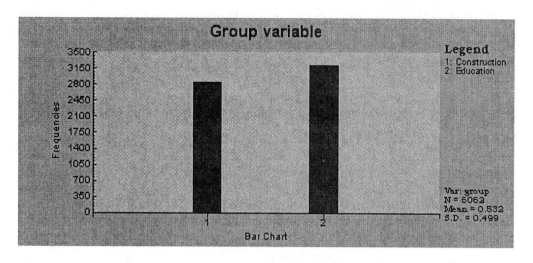

Figure 2.3   Bar Chart of the GROUP Variable in INCOME.PSF

Other demographic variables and their codes are:

GENDER	0 = female; 1 = male
AGE	Age in single years
MARITAL	1 = married; 0 = other
HOURS	Hours worked during last week at all jobs
CITIZEN	1 for native Americans, 0 for all foreign born respondents
INCOME	The natural logarithm of the personal income during 1994
DEGREE	1 for respondents with master's degrees, professional school degree, or doctoral degree; 0 otherwise

Respondents were from 9 regions of the USA, and the state of residence was also given. The variables REGION and STATE represent this information. The full description of the regions and states within regions is presented in Table 2.2. On the respondent level, the variable PERSON is a respondent identity number.

Finally, the variable CONSTANT denotes the intercept term and has a value of 1 for all respondents. The variable INCOME will be used as response variable in all analyses.

2.5 MULTILEVEL EXAMPLES

Table 2.2 Region and State codes

Region	State Code	State Name
New England region (REGION = 1)	11	Maine
	12	New Hampshire
	13	Vermont
	14	Massachusetts
	15	Rhode Island
	16	Connecticut
Middle Atlantic region (REGION=2)	21	New York
	22	New Jersey
	23	Pennsylvania
East North Central region (REGION=3)	31	Ohio
	32	Indiana
	33	Illinois
	34	Michigan
	35	Wisconsin
West North Central region (REGION=4)	41	Minnesota
	42	Iowa
	43	Missouri
	44	North Dakota
	45	South Dakota
	46	Nebraska
	47	Kansas
South Atlantic region (REGION=5)	51	Delaware
	52	Maryland
	53	District of Columbia
	54	Virginia
	55	West Virginia
	56	North Carolina
	57	South Carolina
	58	Georgia
	59	Florida
East South Central region (REGION=6)	61	Kentucky
	62	Tennessee
	63	Alabama
	64	Mississippi
West South Central region (REGION=7)	71	Arkansas
	72	Louisiana
	73	Oklahoma
	74	Texas
Mountain region (REGION=8)	81	Montana
	82	Idaho
	83	Wyoming
	84	Colorado
	85	New Mexico
	86	Arizona
	87	Utah
	88	Nevada
Pacific region (REGION=9)	91	Washington
	92	Oregon
	93	California
	94	Alaska
	95	Hawaii

The models considered here are:

- A 3-level model for the combined group, using INCOME.PSF
- A similar model for the education sector only, using a subset of the data
- A similar model for the construction sector only, using a subset of the data

3-level model for subset of CPC survey data

The data set used, as described in the previous section, is contained in the PSF file INCOME.PSF. The variable labels (first two lines) and the first fifteen data records of this file are shown below.

```
REGION    STATE      AGE   GENDER  MARITAL   HOURS
CITIZEN  PERSON CONSTANT   DEGREE    GROUP  INCOME
1 11 30   1  1 40   1   14751 1   0  0  9.6291
1 11 41   1  1 40   1   14768 1   0  0 10.6041
1 11 46   0  1 36   1   14781 1   0  0 10.1897
1 11 30   1  1 40   1   14813 1   0  0 10.9618
1 11 46   1  1 10   0   14825 1   0  0  9.8009
1 11 63   1  1 25   1   14830 1   0  0 10.2071
1 11 38   1  0 75   1   14833 1   0  0 10.0587
1 11 33   1  1 40   1   14861 1   0  0 10.2400
1 11 25   1  0 60   1   14890 1   0  0 10.2324
1 11 43   1  0 40   1   14949 1   0  0 10.0648
1 11 34   1  1  6   1   14984 1   0  0 10.8208
1 11 46   1  1  0   1   14999 1   0  0  9.6803
1 11 41   1  1 39   1   15009 1   0  0  9.9998
1 11 61   1  1 32   1   15012 1   0  0  9.5468
1 11 51   1  1 25   1   15068 1   0  0 10.4913
```

Again, we start creating the input file by accepting the defaults for the maximum number of iterations, the convergence criterion, and the output options, then provide the title for the analysis (optional).

As all respondents are nested within state of residence, and states are in turn nested within the nine regions, we select REGION as the variable for the level-3 identification variable field. The variables STATE and PERSON are selected as level-2 and level-1 identification variables, respectively.

2.5 MULTILEVEL EXAMPLES

Next, we select INCOME, representing the natural logarithm of personal income, as the response variable for this analysis. The variables AGE, GENDER, MARITAL, HOURS, CITIZEN, CONSTANT, DEGREE, and GROUP are all entered into the model as fixed effects.

Finally, the variable CONSTANT, representing the intercept term, is identified as a random effect on all levels of the hierarchy.

As a result, the input file (INCOME.PR2) looks like this:

```
OPTIONS ;
TITLE=Analysis of CPC data: combined group;
SY=K:\LISREL83\MLEVELEX\INCOME.PSF;
ID1=person;
ID2=state;
ID3=region;
RESPONSE=income;
FIXED=age gender marital hours citizen constant degree group;
RANDOM1=constant;
RANDOM2=constant;
RANDOM3=constant;
```

The data summary and output for the final iteration are given below.

```
              +--------------+
              | DATA SUMMARY |
              +--------------+

NUMBER OF LEVEL 3 UNITS :      9
NUMBER OF LEVEL 2 UNITS :     51
NUMBER OF LEVEL 1 UNITS :   6062

ID3 :    1     2     3     4     5     6     7     8     9
N2  :    6     3     5     7     9     4     4     8     5
N1  :  545   862   785   521  1095   291   598   704   661
```

[Output omitted]

```
ITERATION NUMBER    3

              +----------------------+
              |  FIXED PART OF MODEL |
              +----------------------+
```

COEFFICIENTS	BETA-HAT	STD.ERR.	Z-VALUE	PR > |z|
age	0.01636	0.00101	16.17417	0.00000
gender	0.23710	0.02853	8.31169	0.00000
marital	0.08456	0.02243	3.76975	0.00016
hours	0.01344	0.00065	20.63542	0.00000
citizen	0.28652	0.03449	8.30714	0.00000
constant	8.19488	0.06867	119.34530	0.00000
degree	0.41226	0.02846	14.48697	0.00000
group	0.19798	0.03135	6.31519	0.00000

+----------------------+
| -2 LOG-LIKELIHOOD |
+----------------------+

-2 LOG-LIKELIHOOD = 14222.6158872357

+----------------------+
| RANDOM PART OF MODEL |
+----------------------+

LEVEL 3	TAU-HAT	STD.ERR.	Z-VALUE	PR > |z|
constant/constant	0.00783	0.00472	1.65774	0.09737

LEVEL 2	TAU-HAT	STD.ERR.	Z-VALUE	PR > |z|
constant/constant	0.00522	0.00250	2.08936	0.03668

LEVEL 1	TAU-HAT	STD.ERR.	Z-VALUE	PR > |z|
constant/constant	0.60688	0.01107	54.83990	0.00000

LEVEL 3 COVARIANCE MATRIX

	constant
constant	0.00783

LEVEL 3 CORRELATION MATRIX

	constant
constant	1.0000

2.5 MULTILEVEL EXAMPLES

```
               LEVEL 2 COVARIANCE MATRIX

                     constant

constant        0.00522


               LEVEL 2 CORRELATION MATRIX

                     constant

constant        1.0000


               LEVEL 1 COVARIANCE MATRIX

                     constant

constant        0.6069


               LEVEL 1 CORRELATION MATRIX

                     constant

constant        1.0000
```

CONVERGENCE REACHED IN 3 ITERATIONS

From the output given above, we see that:

- The nine regions had between 291 and 1095 respondents, nested within states. The smallest number of level-2 units within a level-3 unit was 3, for the middle Atlantic region which included only New York, New Jersey, and Pennsylvania.
- All the fixed effects were highly significant. The coefficient for the variable CONSTANT, representing the mean income, was 8.19488. Since the response variable is the natural logarithm of a respondent's annual income, this number translates to a mean income of

$$\exp(8.19488 + 21(0.01636) + 40(0.01344)) = \$8,743$$

for a respondent from the construction sector who is 21 years of age, working 40 hours per week, unmarried, without a higher degree, and not a USA citizen. Although the size of the coefficients is quite small, it should be kept in mind that the natural logarithm of income is used as response variable. The relatively large positive coefficients for GENDER (0.23710), CITIZEN (0.28652), and DEGREE (0.41226) indicate that males, citizens of the USA, and respondents with a high education level tend to earn more when other variables are held constant. A comparison of two respondents with different demographic profiles as given below illustrates this point.

Respondent 1	Respondent 2
AGE=30	AGE=30
HOURS=40	HOURS=40
GROUP=1	GROUP=1
MARITAL=0	MARITAL=0
GENDER=0	GENDER=1
CITIZEN=0	CITIZEN=1
DEGREE=0	DEGREE=1

The first respondent's expected income is calculated as

Expected income =

$$\exp[8.19488 + 30(0.01636) + 40(0.01344) + 0.19798] =$$

$$\exp[9.42126] = \$12,348$$

while the expected income of the second respondent is

Expected income =

$$\exp[8.19488 + 30(0.01636) + 40(0.01344) + 0.19798 + 0.23710 +$$

$$0.28652 + 0.41226] = \exp[10.35714] = \$31,481$$

❑ Income varies most over the respondents (level-1 units), and least over the states (level-2) units as we can see from the variances at these levels, given as 0.60688 and 0.00522, respectively.

In order to take a closer look at the relationships within the construction and educational sectors, two separate data sets are created for these groups and similar models fitted in the next two examples. In the next section, a model for respondents from the education sector will be considered.

2.5 MULTILEVEL EXAMPLES

Three-level model for the educational sector

In the previous example, a 3-level model for the combined education and construction sector respondents from the 1995 CPC survey data was considered. In order to study effects for the educational sector only, a subset of the data in the file INCOME.PSF is used.

We select respondents belonging to the educational sector by using the PRELIS SC (select cases) command and select only those cases for which GROUP=1. The new dataset is saved as EDUC.PSF.

```
CREATE A SUBSET OF THE FULL DATASET
SY=L:\LISREL83\MLEVELEX\INCOME.PSF
SC GROUP=1
OU XM RA=EDUC.PSF
```

The input file for the analysis is exactly the same as in the previous example, with one exception — the variable GROUP is not included as a fixed effect, as this variable now has the value 1 for all respondents in the data.

The input file for this analysis (EDUC.PR2) is shown below.

```
OPTIONS OLS=YES CONVERGE=0.001000 MAXITER=10 OUTPUT=STANDARD ;
TITLE=Analysis of CPC data: education sector group;
SY=L:\LISREL83\MLEVELEX\EDUC.PSF;
ID1=person;
ID2=state;
ID3=region;
RESPONSE=income;
FIXED=age gender marital hours citizen constant degree;
RANDOM1=constant;
RANDOM2=constant;
RANDOM3=constant;
```

Partial output for this analysis follows.

```
ITERATION NUMBER      4

                  +------------------------+
                  |   FIXED PART OF MODEL  |
                  +------------------------+
```

COEFFICIENTS	BETA-HAT	STD.ERR.	Z-VALUE	PR > |Z|
age	0.02001	0.00129	15.49123	0.00000
gender	0.20943	0.02759	7.58961	0.00000
marital	-0.01506	0.02851	-0.52812	0.59741
hours	0.01458	0.00079	18.55257	0.00000
citizen	0.17746	0.05042	3.51950	0.00043
constant	8.38120	0.07850	106.76220	0.00000
degree	0.39622	0.02693	14.71524	0.00000

+------------------------+
| -2 LOG-LIKELIHOOD |
+------------------------+

-2 LOG-LIKELIHOOD = 6991.19474301050

+------------------------+
| RANDOM PART OF MODEL |
+------------------------+

LEVEL 3	TAU-HAT	STD.ERR.	Z-VALUE	PR > |Z|
constant/constant	0.00502	0.00445	1.12840	0.25915

LEVEL 2	TAU-HAT	STD.ERR.	Z-VALUE	PR > |Z|
constant/constant	0.01283	0.00498	2.57744	0.00995

LEVEL 1	TAU-HAT	STD.ERR.	Z-VALUE	PR > |Z|
constant/constant	0.50458	0.01267	39.83364	0.00000

LEVEL 3 COVARIANCE MATRIX

	constant
constant	0.00502

LEVEL 3 CORRELATION MATRIX

	constant
constant	1.0000

2.5 MULTILEVEL EXAMPLES

```
        LEVEL 2 COVARIANCE MATRIX

              constant

constant      0.01283


        LEVEL 2 CORRELATION MATRIX

              constant

constant      1.0000


        LEVEL 1 COVARIANCE MATRIX

              constant

constant      0.5046


        LEVEL 1 CORRELATION MATRIX

              constant

constant      1.0000
```

CONVERGENCE REACHED IN 4 ITERATIONS

For the education sector, the following results are obtained.

- All fixed effects are highly significant and positive, with the exception of the coefficient for marital status (MARITAL). For this group, the coefficient for the intercept (CONSTANT) is 8.38120. From the results of the previous analysis, the intercept for the group of respondents with GROUP=1 was $8.19488 + 0.19798 = 8.39286$, with all other variables held constant. In general, the same trends are observed for the combined and educational sector only groups: larger positive coefficients are obtained for the variables GENDER, CITIZEN, and DEGREE. Using the same two respondent profiles as in the previous example, with the exception of the GROUP variable which was not included in this analysis, we calculate the expected incomes of the two repondents.

	Respondent 1	Respondent 2
	AGE=30	AGE=30
	HOURS=40	HOURS=40
	GROUP=1	GROUP=1
	MARITAL=0	MARITAL=0
	GENDER=0	GENDER=1
	CITIZEN=0	CITIZEN=1
	DEGREE=0	DEGREE=1

The first respondent's expected income is calculated as

Expected income =

$$\exp[8.38120+30(0.02001)+40(0.01458)] = \exp[9.5647] = \$14,252$$

while the expected income of the second respondent is

Expected income =

$$\exp[8.38120+30(0.02001)+40(0.01458)+0.20943+0.17746+0.39622] =$$

$$\exp[10.34781] = \$31,118$$

The difference between the expected income of these respondents is slightly smaller when only the educational sector is considered.

❑ For this sector, the mean income varies little over the nine regions. The variation at level 3 of the model is smaller than for the combined model (0.00783 versus 0.00502) and is not significant at any commonly used level of significance. The conclusion may be reached that most of the variation previously observed at a region level (level 3) was due to differences between the two sectors. Variation at levels 1 and 2 remained significant.

In the last example, we will consider a similar model for the construction sector only.

2.5 MULTILEVEL EXAMPLES

Three-level model for the construction sector

In the previous two examples, a model for the combined education and construction sectors and a model for the educational sector only were fitted to the 1995 CPC survey data. As a final example, we consider a separate model for those respondents active in the construction sector during 1994.

In order to study effects for the construction sector only, a subset of the data in the file INCOME.PSF is used. As before, we select respondents belonging to the construction sector by running a small PRELIS input file using the select cases (SC) command:

```
CREATE A SUBSET OF THE FULL DATASET
SY=L:\LISREL83\MLEVELEX\INCOME.PSF
SC GROUP=0
OU XM RA=CONS.PSF
```

The resulting file CONS.PSF contains only those cases with GROUP=0.

The input file is exactly the same as in the previous example, with one exception — the variable GROUP is not included as a fixed effect, as this variable now has the value 0 for all respondents in the data set CONS.PSF.

The input file (CONS.PR2) for this analysis is shown below.

```
OPTIONS OLS=YES CONVERGE=0.001000 MAXITER=10 OUTPUT=STANDARD ;
TITLE=Analysis of CPC data: construction sector only;
SY=L:\LISREL83\MLEVELEX\CONS.PSF;
ID1=person;
ID2=state;
ID3=region;
RESPONSE=income;
FIXED=age gender marital hours citizen constant degree;
RANDOM1=constant;
RANDOM2=constant;
RANDOM3=constant;
```

Partial output for this analysis is given below.

ITERATION NUMBER   6

+----------------------+
| FIXED PART OF MODEL |
+----------------------+

COEFFICIENTS	BETA-HAT	STD.ERR.	Z-VALUE	PR > |Z|
age	0.01208	0.00157	7.69715	0.00000
gender	0.39847	0.09723	4.09819	0.00004
marital	0.20337	0.03505	5.80180	0.00000
hours	0.01183	0.00110	10.73244	0.00000
citizen	0.32688	0.04805	6.80230	0.00000
constant	8.15061	0.13001	62.69199	0.00000
degree	0.21725	0.22626	0.96018	0.33697

+----------------------+
| -2 LOG-LIKELIHOOD |
+----------------------+

-2 LOG-LIKELIHOOD =   7087.74890333658

+----------------------+
| RANDOM PART OF MODEL |
+----------------------+

LEVEL 3	TAU-HAT	STD.ERR.	Z-VALUE	PR > |Z|
constant/constant	0.01203	0.00747	1.61013	0.10737

LEVEL 2	TAU-HAT	STD.ERR.	Z-VALUE	PR > |Z|
constant/constant	0.00486	0.00408	1.18984	0.23411

LEVEL 1	TAU-HAT	STD.ERR.	Z-VALUE	PR > |Z|
constant/constant	0.70303	0.01880	37.39738	0.00000

LEVEL 3 COVARIANCE MATRIX

           constant

constant   0.01203

2.5 MULTILEVEL EXAMPLES

```
            LEVEL 3 CORRELATION MATRIX

                    constant

constant      1.0000


            LEVEL 2 COVARIANCE MATRIX

                    constant

constant      0.00486


            LEVEL 2 CORRELATION MATRIX

                    constant

constant      1.0000


            LEVEL 1 COVARIANCE MATRIX

                    constant

constant      0.7030


            LEVEL 1 CORRELATION MATRIX

                    constant

constant      1.0000
```

CONVERGENCE REACHED IN 6 ITERATIONS

The following conclusions may be reached from the output given above:

- When the fixed effects for this model is compared to those obtained for the education sector, the coefficients for AGE and HOURS are smaller. The age of a respondent in the construction sector and the number of hours worked will result in a smaller expected increase in annual personal income. In contrast with the education sector, where the effect of marital status (MARITAL) was not significant, a respondent in the construction sector is likely to earn more when the respondent is married, with all other variables held constant.

- The coefficient for CITIZEN is approximately twice that of a respondent from the education sector (0.32688 versus 0.17746). The mean income, with all other variables held fixed at 0, is 8.15061 (as natural logarithm). With all other variables held constant, this translates into a $1,339 difference in baseline income between citizens and non-citizens in the construction sector. The baseline expected income for a US citizen working in the construction sector can be calculated as

Expected baseline =

$$\exp(8.15061 + 0.32688) = \$4,805$$

For a US citizen in the education sector, the expected baseline income is calculated as

Expected baseline =

$$\exp(8.38120 + 0.17746) = \$5,211$$

- Again, the largest coefficients obtained are for GENDER, CITIZEN, and DEGREE. Where the coefficient for GENDER was 0.20943 in the education sector, the coefficient for the construction sector is approximately twice that, at 0.39847. From the output above, it is seen that the coefficient for DEGREE is not significant. A closer examination of the data, using PRELIS data screening features, reveals that only 14 respondents have a high level of education (masters, professional, or PhD degree). If the same model is fitted without the degree predictor, all other estimated parameter values basically remain unchanged.

- Turning to the random effects, we see that the only significant variation is over respondents. At both state and division level, the variation is not significant. From this, combined with the results of the analysis for the education sector (see p. 116), we conclude that the significant variation at level 2 seen in the combined model (see p. 121), is probably due to the differences between these two groups.

3 Other New Statistical Features

This chapter[1] presents and illustrates all new statistical features in LIS-REL 8.30. These include, in particular, two-stage least-squares estimation, exploratory factor analysis, principal components, normal scores, and latent variable scores.

Most of the recent developments in structural equation modeling have focused on full information methods such as maximum likelihood (ML) or asymptotically distribution free methods (ADF or WLS) and a discussion of the best parameter estimates and standard errors. While such methods are statistically optimal in theory under certain assumptions, such optimality is seldom required in practice either because the model is only tentative or there may be other sources of error in the data of far more importance than the statistical errors, see, *e.g.*, Jöreskog (1993). In such situations, *two-stage least-squares estimates* (TSLS) using instrumental variables may be useful. TSLS estimates and their standard errors may be obtained quickly without iterations for several typical structural equation models useful in the early stages of investigation. They often provide sufficient information to judge whether the model is reasonable or not.

Exploratory factor analysis is useful in the early stages of investigation to study the measurement properties of the observed variables. LISREL 8.30 provides ML estimates of factor loadings (unrotated, varimax-rotated, or promax-rotated) and TSLS estimates of a rotation to a reference variables solution. The latter is particularly useful because it gives standard errors of the factor loadings. The number of factors may be specified by the user or will be determined automatically by the program. Factor scores for the

---

[1]Written by Karl Jöreskog and Dag Sörbom, Uppsala University, Department of Information Science, PO Box 513, SE–75120 Uppsala, Sweden

factors in the reference variables solution may be obtained and saved in a file for further use.

Principal components and principal components scores can be obtained for all components or for any specified number of components.

Normal scores may be computed for ordinal and continuous variables. For continuous variables these give the possibility of effectively normalizing non-normal variables. Normal scores provide a way to deal with non-normality in small and moderate sample sizes. Recall that ML requires multivariate normality and other methods (such as ADF) require large or huge samples to deal with non-normality.

Latent variable scores are individual scores on the latent variables in an estimated structural equation model. These can be obtained for any estimated single group LISREL model. One possible use of such scores is to investigate various structural models after the measurement models for the observed variables have been well established.

3.1 Two-Stage Least-Squares

Suppose we want to estimate the linear relationship between a dependent variable y and a set of explanatory variables $\mathbf{x}' = (x_1, x_2, \ldots, x_p)$:

$$y = \gamma_1 x_1 + \gamma_2 x_2 + \cdots + \gamma_p x_p + u , \qquad (3.1)$$

or in matrix form

$$y = \boldsymbol{\gamma}' \mathbf{x} + u , \qquad (3.2)$$

where u is a random error term and $\boldsymbol{\gamma}' = (\gamma_1, \gamma_2, \ldots, \gamma_p)$ is a vector of coefficients to be estimated. For simplicity of the argument, we assume that all variables are measured in deviations from their means so that there is no intercept in (3.1) or (3.2). It will be shown later how to estimate a relationship with an intercept term.

If u is uncorrelated with $x_1, \ldots, x_p$, ordinary least-squares (OLS) can be used to obtain a consistent estimate of $\boldsymbol{\gamma}$, yielding the wellknown solution

$$\hat{\boldsymbol{\gamma}} = \mathbf{S}_{xx}^{-1} \mathbf{s}_{xy} , \qquad (3.3)$$

where $\hat{\gamma}$ is a $p \times 1$ vector of estimated γ's, $\mathbf{S}_{xx}$ is the $p \times p$ sample covariance matrix of the x-variables and $\mathbf{s}_{xy}$ is the $p \times 1$ vector of sample covariances between the x-variables and y.

If u is *correlated* with one or more of the x_i, however, the OLS estimate in (3.3) is not consistent, *i.e.*, it is biased even in large samples. The bias can be positive or negative, large or small depending on the correlations between u and $\mathbf{x}$. But suppose some *instrumental variables* $\mathbf{z}' = (z_1, \ldots, z_q)$ are available, where $q \geq p$. An instrumental variable is a variable which is uncorrelated with u. Then the following two-stage least-squares (TSLS) estimator:

$$\hat{\gamma} = (\mathbf{S}'_{zx} \mathbf{S}_{zz}^{-1} \mathbf{S}_{zx})^{-1} \mathbf{S}'_{zx} \mathbf{S}_{zz}^{-1} \mathbf{s}_{zy} , \qquad (3.4)$$

can be used to estimate γ consistently, where $\mathbf{S}_{zx}$ is the $q \times p$ matrix of sample covariances between the z-variables and the x-variables, $\mathbf{S}_{zz}$ is the $p \times p$ sample covariance matrix of the z-variables, and $\mathbf{s}_{zy}$ is the $q \times 1$ vector of sample covariances between the z-variables and y.

The usual way of deriving (3.4) is as a two-step procedure, see, *e.g.*, Goldberger (1964) or Theil (1971):

Step 1 Estimate the OLS regression of each x_i on $\mathbf{z}$, yielding $\mathbf{x} = \hat{\mathbf{B}}\mathbf{z} + \mathbf{v}$, where $\hat{\mathbf{B}} = \mathbf{S}'_{zx} \mathbf{S}_{zz}^{-1}$.

Step 2 Replace $\mathbf{x}$ in (3.2) with $\hat{\mathbf{x}} = \hat{\mathbf{B}}\mathbf{z}$ and estimate the OLS regression of y on $\hat{\mathbf{x}}$. Combining the results of Steps 1 and 2 gives (3.4).

However, (3.4) shows that this estimator can be computed in one step once the covariance matrix of the variables involved has been computed. It has the advantage that it does not require any distributional assumptions. It is valid if the estimated relationship is linear, the required variances and covariances exist, and the inverses in (3.4) exist.

The covariance matrix of $\hat{\gamma}$ can be estimated as

$$(n-p)^{-1} \hat{\sigma}_{uu} (\mathbf{S}'_{zx} \mathbf{S}_{zz}^{-1} \mathbf{S}_{zx})^{-1} , \qquad (3.5)$$

where

$$\hat{\sigma}_{uu} = s_{yy} - 2\hat{\gamma}' \mathbf{s}_{xy} + \hat{\gamma}' \mathbf{S}_{xx} \hat{\gamma} \qquad (3.6)$$

is a consistent estimate of the variance of u and $n = N - 1$, where N is the sample size. The standard errors of the estimated γ's are the square roots of the diagonal elements of (3.5).

Every x-variable which is uncorrelated with u may serve as an instrumental variable z. If all x-variables are uncorrelated with u, the x-variables themselves serve as instrumental variables. It may be easily verified that if $\mathbf{z} = \mathbf{x}$, then (3.4) reduces to (3.3) and (3.5) reduces to the well-known OLS formula

$$(n - p)^{-1} \hat{\sigma}_{uu} \mathbf{S}_{xx}^{-1} .$$

For every x-variable which is correlated with u there must be at least one instrumental variable outside the set of x-variables. Usually there is exactly one instrumental variable for each x-variable which is correlated with u, but it is possible to use more than one.

3.1.1 Implementation

With PRELIS 2.20 or earlier one could do univariate or multivariate regression, including ANOVA, ANCOVA, MANOVA, and MANCOVA and other variations of the general multivariate linear model, with the PRELIS command

```
RG Y-Varlist ON X-Varlist
```

Two-stage least-squares (TSLS) can be seen as a simple extension of this command. With LISREL 8.30, one can do TSLS estimation in PRELIS or in LISREL with either the LISREL or the SIMPLIS command language. With PRELIS one can read in raw data on any number of variables and compute the covariance matrix after elimination of missing values,[2] if any. PRELIS is particularly convenient when the instrumental variables are constructed as functions of other variables in the data set, see *Example: The Kenny–Judd Model* in Section 3.7. With LISREL one can read a covariance matrix for any number of variables. In either case, the program will automatically select the subsets of y-, x-, and z-variables needed from an RG command.

---

[2]One can also impute missing values, recode variables and do various transformations of the variables.

In PRELIS and in the LISREL command language the RG command is

```
RG Y-Variable ON X-Varlist WITH Z-Varlist
```

where Y-Variable, X-Varlist, and Z-Varlist are lists of names of variables, corresponding to the y-, x-, and z-variables, respectively. This command is interpreted by the program as "Estimate the relationship between the y-variable in Y-Variable, and the x-variables in X-Varlist, using the z-variables in Z-Varlist as instrumental variables." The word ON must follow the name of the y-variable and the word WITH separates the x-variables from the z-variables. If the part WITH Z-Varlist of the RG command is omitted, the program will set z=x, *i.e.*, it will perform an OLS regression. Otherwise, if the part WITH Z-Varlist of the RG command is included, there must be at least as many variables in Z-Varlist as in X-Varlist.

In the SIMPLIS command language, simply use the command

```
Regress Y-Variable On X-Varlist With Z-Varlist as Instrumental Variables
```

Here also, the words On and With are important words with the same meaning as before.

Although PRELIS and LISREL will accept more than one variable before the word ON, it is seldom possible to use the same instrumental variables in different equations. An exception is for multivariate regression models, as our first example illustrates.

3.1.2 Residual Analysis

The estimated residual $\hat{u} = y - \hat{\gamma}'\mathbf{x}$ may be computed for each individual in the sample by including RES=varname at the end of the RG command in PRELIS, where varname is the name the user assigns for the residual. After the regression equation has been estimated, this variable may be augmented to the raw data and used in another PRELIS run. This is illustrated in Section 3.7. Once the residual is available in the raw data, one can examine its behavior in the sample by importing the raw data into PRELIS and viewing various univariate and bivariate graphs.

If $q = p$ and $\mathbf{S}_{zx}$ is nonsingular, the residual $\hat{u}$ is uncorrelated with the instrumental variables $\mathbf{z}$, just as the residual in ordinary least squares (OLS) is uncorrelated with the explanatory variables $\mathbf{x}$. If $q > p$ and the rank of $\mathbf{S}_{zx}$ is p, there will be p linear combinations of $\mathbf{z}$ that are uncorrelated with the residual $\hat{u}$.

3.1.3 Regression Models

A regression model is a model of the form

$$\mathbf{y} = \boldsymbol{\alpha} + \boldsymbol{\Gamma}\mathbf{x} + \mathbf{u} , \qquad (3.7)$$

where $\mathbf{y}' = (y_1, y_2, \ldots, y_p)$ is a set of jointly dependent variables, $\mathbf{x}' = (x_1, x_2, \ldots, x_q)$ is a set of explanatory variables uncorrelated with the error terms $\mathbf{u}' = (u_1, u_2, \ldots, u_p)$, $\boldsymbol{\alpha}$ is a vector of intercept terms and $\boldsymbol{\Gamma}$ is a matrix of regression parameters.

Example: Prediction of Grade Averages

The following example illustrates the case of two dependent variables y_1 and y_2 and three explanatory variables x_1, x_2, and x_3.

Finn (1974) presents the data given in Table 3.1. These data (given in file GRAV.RAW) represent the scores of fifteen freshmen at a large midwestern university on five educational measures. The five measures are

- $y_1 = $ grade average for required courses taken (GRAVEREQ)
- $y_2 = $ grade average for elective courses taken (GRAVELEC)
- $x_1 = $ high-school general knowledge test, taken previous year (KNOWLEDG)
- $x_2 = $ IQ score from previous year (IQPREVYR)
- $x_3 = $ educational motivation score from previous year (ED MOTIV)

We examine the predictive value of x_1, x_2 and x_3 in predicting the grade averages y_1 and y_2.

To estimate the regression of y_1 and y_2 on x_1, x_2, and x_3 use the following PRELIS command file (see file GRAV.PR2).

3.1 TWO-STAGE LEAST-SQUARES

Table 3.1
Scores for Fifteen College Freshmen on Five Educational Measures

Case	y_1	y_2	x_1	x_2	x_3
1	.8	2.0	72	114	17.3
2	2.2	2.2	78	117	17.6
3	1.6	2.0	84	117	15.0
4	2.6	3.7	95	120	18.0
5	2.7	3.2	88	117	18.7
6	2.1	3.2	83	123	17.9
7	3.1	3.7	92	118	17.3
8	3.0	3.1	86	114	18.1
9	3.2	2.6	88	114	16.0
10	2.6	3.2	80	115	16.4
11	2.7	2.8	87	114	17.6
12	3.0	2.4	94	112	19.5
13	1.6	1.4	73	115	12.7
14	.9	1.0	80	111	17.0
15	1.9	1.2	83	112	16.1

```
Prediction of Grade Averages
DATA NI=5
LABELS; GRAVEREQ GRAVELEC KNOWLEDG IQPREVYR 'ED MOTIV'
RAWDATA=GRAV.RAW
CONTINUOUS ALL
RG GRAVEREQ GRAVELEC ON KNOWLEDG IQPREVYR 'ED MOTIV'
OUTPUT
```

The result is given in the output as

```
Estimated Equations

GRAVEREQ =  - 5.619 + 0.0854*KNOWLEDG + 0.00822*IQPREVYR - 0.0149*ED MOTIV
            (5.614) (0.0270)            (0.0485)            (0.112)
            -1.001   3.168                0.169              -0.134

            + Error, R² = 0.568

Error Variance = 0.327
```

```
GRAVELEC =  - 20.405 + 0.0472*KNOWLEDG + 0.145*IQPREVYR + 0.126*ED MOTIV
             (5.398)   (0.0259)           (0.0467)        (0.107)
             -3.780    1.823              3.117           1.170

          + Error, R² = 0.685

Error Variance = 0.302
```

3.1.4 Econometric Models

Two-stage least-squares (TSLS) is particularly useful for estimating econometric models of the form

$$\mathbf{y} = \mathbf{B}\mathbf{y} + \mathbf{\Gamma}\mathbf{x} + \mathbf{u} , \qquad (3.8)$$

where $\mathbf{y}' = (y_1, y_2, \ldots, y_p)$ is a set of endogenous or jointly dependent variables, $\mathbf{x}' = (x_1, x_2, \ldots, x_q)$ is a set of exogenous or predetermined variables uncorrelated with the error terms $\mathbf{u}' = (u_1, u_2, \ldots, u_p)$, and $\mathbf{B}$ and $\mathbf{\Gamma}$ are parameter matrices. In a typical econometric model (3.8) represents an *interdependent system* or a *non-recursive system* in which the y-variables cannot be ordered so that $\mathbf{B}$ is sub-diagonal. A typical feature of such a model is that not all y-variables and not all x-variables are included in each equation. The i-th equation in (3.8) is of the form

$$y_i = \boldsymbol{\beta}_{(i)} \mathbf{y}_{(i)} + \boldsymbol{\gamma}_{(i)} \mathbf{x}_{(i)} + u_i , \qquad (3.9)$$

where $\mathbf{y}_{(i)}$ is a vector of the y-variables included in the right side of the i-th equation and $\mathbf{x}_{(i)}$ is a vector of the x-variables included in the i-th equation and $\boldsymbol{\beta}_{(i)}$ and $\boldsymbol{\gamma}_{(i)}$ are row vectors formed from the non-zero elements of $\mathbf{B}$ and $\mathbf{\Gamma}$, respectively.

A necessary condition for identification of each equation (3.9) is that the number of x-variables excluded from the equation is at least as great as the number of y-variables included in the right side of the equation. This is the so-called *order condition*. In other words, for every y-variable on the right side of an equation there must be at least one x-variable excluded from that equation. There is also a sufficient condition for identification, the so-called *rank condition*, but this is often difficult to apply in practice. For further information on identification of interdependent systems, see, *e.g.*, Goldberger (1964, pp. 313–318).

3.1 TWO-STAGE LEAST-SQUARES

Example: Klein's Model I of US Economy

Klein's (1950) Model I is a classical econometric model which has been used extensively as a benchmark problem for studying econometric methods. It is an eight-equation system based on annual data for the United States in the period between the two World Wars. It is dynamic in the sense that elements of time play important roles in the model. The three behavioral equations of Klein's Model are

$$C_t = a_1 P_t + a_2 P_{t-1} + a_3 W_t + u_1$$
$$I_t = b_1 P_t + b_2 P_{t-1} + b_3 K_{t-1} + u_2$$
$$W_t^* = c_1 E_t + c_2 E_{t-1} + c_3 A_t + u_3$$

In addition to these stochastic equations, the model includes five identities (definitional equations):

$$P_t = Y_t - W_t$$
$$Y_t = C_t + I_t + G_t - T_t$$
$$K_t = K_{t-1} + I_t$$
$$W_t = W_t^* + W_t^{**}$$
$$E_t = Y_t + T_t - W_t^{**}$$

The endogenous variables are

C_t = Aggregate Consumption (y_1)
I_t = Net Investment (y_2)
W_t^* = Private Wage Bill (y_3)
P_t = Total Profits (y_4)
Y_t = Total Income (y_5)
K_t = End-of-Year Capital Stock (y_6)
W_t = Total Wage Bill (y_7)
E_t = Total Production of Private Industry (y_8)

The predetermined variables are the exogenous variables

$$W_t^{**} = \text{Government Wage Bill } (x_1)$$
$$T_t = \text{Taxes } (x_2)$$
$$G_t = \text{Government Non-Wage Expenditures } (x_3)$$
$$A_t = \text{Time in Years From 1931 } (x_4)$$

and the lagged endogenous variables P_{t-1} (x_5), K_{t-1} (x_6), and E_{t-1} (x_7). All variables except A_t are in billions of 1934 dollars. Annual time series data for 1921–1941 are given in Table 3.2, which has been computed from Theil's (1971) Table 9.1. These data are given in file KLEIN.RAW.

In the consumption function there are 2 y-variables included on the right side, namely P_t and W_t and 6 x-variables excluded namely W_t^{**}, T_t, G_t, A_t, K_{k-1}, and E_{t-1} so that the order condition is fulfilled. Similarly, it can be verified that the order condition is met also for the other two equations.

To estimate the consumption function, we use C_t as the y-variable, P_t, P_{t-1}, and W_t as x-variables, and all the predetermined variables as z-variables, in the sense of equation (3.1). A PRELIS command file to do this is (see KLEIN1.PR2):

```
Estimating Klein's Consumption Function
Data NI=15
Labels
C P_1 W* I K_1 E_1 W** T A P K E W Y G
Rawdata=Klein.raw
Continuous All
RG C on P P_1 W with W** T G A P_1 K_1 E_1
Output
```

The estimated consumption equation is given in the output as:[3]

```
Estimated Equations
```
$$C = 16.555 + 0.0173*P + 0.216*P\_1 + 0.810*W + \text{Error}, \quad R^2 = 0.977$$
$$(1.468) \quad (0.131) \quad (0.119) \quad (0.0447)$$

---

[3]These results agree with those given in Goldberger (1964, p. 336) and Theil (1971, p. 458), except that these authors do not give the estimate of the intercept and they divide by N in (3.5) instead of $N - 1 - p$

3.1 TWO-STAGE LEAST-SQUARES

Table 3.2   Time Series Data for Klein's Model I

t	C_t	P_{t-1}	W_t^*	I_t	K_{t-1}	E_{t-1}	W_t^{**}
1921	41.9	12.7	25.5	-0.2	182.8	44.9	2.7
1922	45.0	12.4	29.3	1.9	182.6	45.6	2.9
1923	49.2	16.9	34.1	5.2	184.5	50.1	2.9
1924	50.6	18.4	33.9	3.0	189.7	57.2	3.1
1925	52.6	19.4	35.4	5.1	192.7	57.1	3.2
1926	55.1	20.1	37.4	5.6	197.8	61.0	3.3
1927	56.2	19.6	37.9	4.2	203.4	64.0	3.6
1928	57.3	19.8	39.2	3.0	207.6	64.4	3.7
1929	57.8	21.1	41.3	5.1	210.6	64.5	4.0
1930	55.0	21.7	37.9	1.0	215.7	67.0	4.2
1931	50.9	15.6	34.5	-3.4	216.7	61.2	4.8
1932	45.6	11.4	29.0	-6.2	213.3	53.4	5.3
1933	46.5	7.0	28.5	-5.1	207.1	44.3	5.6
1934	48.7	11.2	30.6	-3.0	202.0	45.1	6.0
1935	51.3	12.3	33.2	-1.3	199.0	49.7	6.1
1936	57.7	14.0	36.8	2.1	197.7	54.4	7.4
1937	58.7	17.6	41.0	2.0	199.8	62.7	6.7
1938	57.5	17.3	38.2	-1.9	201.8	65.0	7.7
1939	61.6	15.3	41.6	1.3	199.9	60.9	7.8
1940	65.0	19.0	45.0	3.3	201.2	69.5	8.0
1941	69.7	21.1	53.3	4.9	204.5	75.7	8.5

t	T_t	A_t	P_t	K_t	E_t	W_t	Y_t	G_t
1921	7.7	-10.0	12.4	182.6	45.6	28.2	40.6	6.6
1922	3.9	-9.0	16.9	184.5	50.1	32.2	49.1	6.1
1923	4.7	-8.0	18.4	189.7	57.2	37.0	55.4	5.7
1924	3.8	-7.0	19.4	192.7	57.1	37.0	56.4	6.6
1925	5.5	-6.0	20.1	197.8	61.0	38.6	58.7	6.5
1926	7.0	-5.0	19.6	203.4	64.0	40.7	60.3	6.6
1927	6.7	-4.0	19.8	207.6	64.4	41.5	61.3	7.6
1928	4.2	-3.0	21.1	210.6	64.5	42.9	64.0	7.9
1929	4.0	-2.0	21.7	215.7	67.0	45.3	67.0	8.1
1930	7.7	-1.0	15.6	216.7	61.2	42.1	57.7	9.4
1931	7.5	0.0	11.4	213.3	53.4	39.3	50.7	10.7
1932	8.3	1.0	7.0	207.1	44.3	34.3	41.3	10.2
1933	5.4	2.0	11.2	202.0	45.1	34.1	45.3	9.3
1934	6.8	3.0	12.3	199.0	49.7	36.6	48.9	10.0
1935	7.2	4.0	14.0	197.7	54.4	39.3	53.3	10.5
1936	8.3	5.0	17.6	199.8	62.7	44.2	61.8	10.3
1937	6.7	6.0	17.3	201.8	65.0	47.7	65.0	11.0
1938	7.4	7.0	15.3	199.9	60.9	45.9	61.2	13.0
1939	8.9	8.0	19.0	201.2	69.5	49.4	68.4	14.4
1940	9.6	9.0	21.1	204.5	75.7	53.0	74.1	15.4
1941	11.6	10.0	23.5	209.4	88.4	61.8	85.3	22.3

```
              11.277    0.132     1.814      18.111
Error Variance = 1.290

Instrumental Variables: W** T G A P_1 K_1 E_1
```

Note that one cannot estimate this equation by regressing C on P_t, P_{t-1}, and W_t, as u_1 is not uncorrelated with P_t and W_t. Instead of

```
RG C on P P_1 W with W** T G A P_1 K_1 E_1
```

one can write

```
Equation C = P P_1 W with W** T G A P_1 K_1 E_1
```

and the word Equation can be abbreviated as Eq. Upper case or lower case can be used.

The intercept term in the equation will always be estimated by PRELIS. To estimate the equation without the intercept term use the SIMPLIS or LISREL command language and read a covariance matrix. To estimate the equation in standardized form, read a correlation matrix or specify MA=KM instead in LISREL.

The following PRELIS command file will add the residual to the variables and compute the covariance matrix of the extended set of variables (see file KLEIN2.PR2):

```
Estimating Klein's Consumption Function
Data NI=15
Labels
C P_1 W* I K_1 E_1 W** T A P K E W Y G
Rawdata=Klein.raw
Continuous All
Equation C = P P_1 W with W** T G A P_1 K_1 E_1 RES=U
Output Matrix=CM
```

One can estimate all three behavioral equations in Klein's Model in one run. The following command file (see file KLEIN3.PR2) will do that and add the three residuals to the raw data in a file KLEINEXT.RAW.

3.1 TWO-STAGE LEAST-SQUARES

```
Estimating Klein's Model I with TSLS
Data NI=15
Labels; C P_1 W* I K_1 E_1 W** T A P K E W Y G
Rawdata=KLEIN.RAW
Continuous All
EQ C  = P P_1 W    with  W** T G A P_1 K_1 E_1 RES=U_1
EQ I  = P P_1 K_1 with  W** T G A P_1 K_1 E_1 RES=U_2
EQ W* = E E_1 A    with  W** T G A P_1 K_1 E_1 RES=U_3
Output Matrix=CM Raw=KLEINEXT.RAW
```

This gives the following result.

Estimated Equations

```
C =  16.555 + 0.0173*P + 0.216*P_1 + 0.810*W + Error, R² = 0.977
    (1.468)  (0.131)   (0.119)    (0.0447)
    11.277   0.132     1.814      18.111

I =  20.278 + 0.150*P + 0.616*P_1 - 0.158*K_1 + Error, R² = 0.885
    (8.383)  (0.193)   (0.181)    (0.0402)
    2.419    0.780     3.404      -3.930

W* = 1.500 + 0.439*E + 0.147*E_1 + 0.130*A + Error, R² = 0.987
    (1.276)  (0.0396) (0.0432)   (0.0324)
    1.176   11.082    3.398      4.026
```

Example: Tintner's Meat Market Model

Tintner (1952, pp. 176–179) formulated a model for the American meat market:

$$y_1 = a_1 y_2 + a_2 x_1 + u_1 , \qquad (3.10)$$

$$y_1 = b_1 y_2 + b_2 x_2 + b_3 x_3 + u_2 , \qquad (3.11)$$

where the dependent variables are

y_1 = meat consumption per capita (pounds)

y_2 = meat price (1935–39 = 100)

and the predetermined variables are

x_1 = disposable income per capita (dollars)

x_2 = unit cost of meat processing (1935–39 = 100)

x_3 = cost of agricultural production (1935–39 = 100)

Note that both equations (3.10) and (3.11) have y_1 on the left side. This is so because (3.10) represents the demand for meat and (3.11) represents the supply of meat and in a free market these are supposed to be equal. Such a model cannot be estimated as a LISREL model.[4] The sample covariance matrix based on annual data for United States 1919–1941 ($N = 23$) is given in Table 3.3.[5]

Table 3.3
Covariance Matrix for Variables in Tintner's Meat Market Model

x_1	x_2	x_3	y_1	y_2
3792.439				
164.169	115.218			
554.762	33.217	119.409		
166.905	-24.385	44.721	62.252	
379.754	38.651	56.171	-16.020	71.886

To estimate the equations (3.10) and (3.11) directly by TSLS one can use the following SIMPLIS command file (see file TINTNER1.SPL)

```
Estimating Tintner's Demand Function
Observed Variables: X1-X3 Y1 Y2
Covariance Matrix from File TINTNER.COV
Sample Size = 23
Regress Y1 on Y2 and X1 with X1-X3 as Instrumental Variables
Regress Y1 on Y2 X2 and X3 with X1-X3 as Instrumental Variables
End of Problem
```

The output gives the two estimated equations as

```
Estimating Tintner's Demand Function

Estimated Equations
```

$$Y1 = -1.58*Y2 + 0.20*X1 + \text{Error}, \quad R^2 = 0.42$$
$$\quad\quad (0.62) \quad\quad (0.066)$$
$$\quad\quad -2.53 \quad\quad\quad 3.05$$

---

[4] To estimate the model as a LISREL model, one would have to rewrite (3.11) as $y_2 = c_1 y_1 + c_2 x_2 + c_3 x_3 + u_3$, where $c_1 = 1/b_1$, $c_2 = -b_2/b_1$, $c_3 = -b_3/b_1$, and $u_3 = -1/b_1$.

[5] This was obtained by dividing the numbers in Goldberger (1964, p. 321) by 22

```
Error Variance = 35.96

Instrumental Variables: X1 X2 X3

    Y1 = - 0.32*Y2 - 0.28*X2 + 0.60*X3 + Error, R² = 0.71
          (0.29)     (0.11)     (0.15)
          -1.10      -2.43       3.96

Error Variance = 18.22

Instrumental Variables: X1 X2 X3
```

Those who want to use a LISREL command file instead of SIMPLIS, can use the following (see file TINTNER2.LS8):

```
Estimating Tintner's Demand and Supply Functions
DA NI=5 NO=23
LA
X1 X2 X3 Y1 Y2
CM
  3792.439
   164.169    115.218
   554.762     33.217    119.409
   166.905    -24.385     44.721     62.252
   379.754     38.651     56.171    -16.020     71.886
RG Y1 ON Y2 X1 WITH X1 X2 X3
RG Y1 ON Y2 X2 X3 WITH X1 X2 X3
OU
```

3.2 Exploratory Factor Analysis

It is important to distinguish between exploratory and confirmatory analysis. In an exploratory analysis, one wants to explore the empirical data to discover and detect characteristic features and interesting relationships without imposing any definite model on the data. An exploratory analysis may be structure generating, model generating, or hypothesis generating. In confirmatory analysis, on the other hand, one builds a model assumed to describe, explain, or account for the empirical data in terms of relatively few parameters. The model is based on *a priori* information about the data structure in the form of a specified theory or hypothesis, a given classificatory design for items or subtests according to objective features of content and format, known experimental conditions, or knowledge from previous studies based on extensive data.

Exploratory factor analysis is a technique often used to detect and assess latent sources of variation and covariation in observed measurements. It is widely recognized that exploratory factor analysis can be quite useful in the early stages of experimentation or test development. Thurstone's (1938) primary mental abilities, French's (1951) factors in aptitude and achievement tests and Guilford's (1956) structure of intelligence are good examples of this. The results of an exploratory factor analysis may have heuristic and suggestive value and may generate hypotheses which are capable of more objective testing by other multivariate methods. As more knowledge is gained about the nature of social and psychological measurements, however, exploratory factor analysis may not be a useful tool and may even become a hindrance.

Most studies are to some extent both exploratory and confirmatory since they involve some variables of known and other variables of unknown composition. The former should be chosen with great care in order that as much information as possible about the latter may be extracted. It is highly desirable that a hypothesis which has been suggested by mainly exploratory procedures should subsequently be confirmed, or disproved, by obtaining new data and subjecting these to more rigorous statistical techniques.

The basic idea of factor analysis is the following. For a given set of response variables $x_1, \ldots, x_p$ one wants to find a set of underlying latent factors $\xi_1, \ldots, \xi_k$, fewer in number than the observed variables. These latent factors are supposed to account for the intercorrelations of the response variables in the sense that when the factors are partialed out from the observed variables, there should no longer remain any correlations between these. If both the observed response variables and the latent facors are measured in deviations from the mean, this leads to the model (see Jöreskog, 1979)

$$x_i = \lambda_{i1}\xi_1 + \lambda_{i2}\xi_2 + \cdots + \lambda_{in}\xi_k + \delta_i , \qquad (3.12)$$

where δ_i, the unique part of x_i, is assumed to be uncorrelated with $\xi_1, \xi_2, \ldots, \xi_k$ and with δ_j for $j \neq i$. The unique part δ_i consists of two components: a specific factor s_i and a pure random measurement error e_i. These are indistinguishable, unless the measurements x_i are designed in such a way that they can be separately identified (panel designs and multitrait-multimethod designs). The term δ_i is often called the *measurement error* in x_i even though it is widely recognized that this term may also contain a specific factor as stated above.

In a confirmatory factor analysis, the investigator has such knowledge about the factorial nature of the variables that he/she is able to specify that each measure x_i depends only on a few of the factors ξ_j. If x_i does not depend on ξ_j, $\lambda_{ij} = 0$ in (3.12). In a path diagram, this means that there is no one-way arrow from ξ_j to x_i. In many applications, the latent factor ξ_j represents a theoretical construct and the observed measures x_i are designed to be indicators of this construct. In this case there is only one non-zero λ_{ij} in each equation (3.12).

Equation (3.12) can be written in matrix form as

$$\mathbf{x} = \mathbf{\Lambda}\boldsymbol{\xi} + \boldsymbol{\delta} , \tag{3.13}$$

where $\mathbf{x}' = (x_1, x_2, \ldots, x_p)$, $\boldsymbol{\xi}' = (\xi_1, \xi_2, \ldots, \xi_k)$, $\boldsymbol{\delta}' = (\delta_1, \delta_2, \ldots, \delta_p)$. The matrix $\mathbf{\Lambda}$ of order $p \times k$ is called the factor matrix or the factor loadings matrix.

The assumption that $\boldsymbol{\delta}$ is uncorrelated with $\boldsymbol{\xi}$, implies that the covariance matrix $\mathbf{\Sigma}$ of $\mathbf{x}$ is

$$\mathbf{\Sigma} = \mathbf{\Lambda}\mathbf{\Phi}\mathbf{\Lambda}' + \mathbf{\Theta} , \tag{3.14}$$

where $\mathbf{\Phi}$ and $\mathbf{\Theta}$ are the covariance matrices of $\boldsymbol{\xi}$ and $\boldsymbol{\delta}$, respectively.

If $k > 1$ and there are no restrictions on $\mathbf{\Lambda}$, i.e., in the exploratory case, the factors $\boldsymbol{\xi}$ are not uniquely defined, because one can make an arbitrary linear transformation of the factors. Let $\mathbf{T}$ be an arbitrary non-singular matrix of order $k \times k$ and let

$$\boldsymbol{\xi}^* = \mathbf{T}\boldsymbol{\xi} \quad \mathbf{\Lambda}^* = \mathbf{\Lambda}\mathbf{T}^{-1} \quad \mathbf{\Phi}^* = \mathbf{T}\mathbf{\Phi}\mathbf{T}'$$

Then we have identically

$$\mathbf{\Lambda}^*\boldsymbol{\xi}^* \equiv \mathbf{\Lambda}\boldsymbol{\xi} \quad \mathbf{\Lambda}^*\mathbf{\Phi}^*\mathbf{\Lambda}^{*'} \equiv \mathbf{\Lambda}\mathbf{\Phi}\mathbf{\Lambda}'$$

This shows that at least k^2 independent conditions must be imposed on Λ and/or Φ to make these identified. It is common to assume that the factors are uncorrelated and standardized, in which case T is restricted to be an orthogonal matrix. Such factors and factor loadings are only determined up to an orthogonal transformation.

Exploratory factor analysis is usually performed in two steps: First, estimate an arbitrary matrix Λ; next, transform this matrix according to external criteria for simple structure to facilitate interpretation of the data. With TSLS estimation, one must use a different approach. If Λ has rank k, one can choose a transformation T such that there will be k rows of Λ that form an identity matrix. For simplicity of exposition, we assume that the variables in x have been ordered so that the first k rows of Λ form an identity matrix. The variables that correspond to the identity matrix are called reference variables and this solution is called a reference variables solution. In practice, it is best to choose as reference variables those variables that are the best indicators of each factor.

Partitioning x into two parts $x_1(k \times 1)$ and $x_2(q \times 1)$, where $q = p - k$, and δ similarly into $\delta_1(k \times 1)$ and $\delta_2(q \times 1)$, (3.13) can be written

$$x_1 = \xi + \delta_1 \qquad (3.15)$$
$$x_2 = \Lambda_2 \xi + \delta_2 , \qquad (3.16)$$

where $\Lambda_2(q \times k)$ consists of the last $q = p - k$ rows of Λ. The matrix Λ_2 may, but need not, contain *a priori* specified elements. We say that the model is *unrestricted* when Λ_2 is entirely unspecified and that the model is *restricted* when Λ_2 contains *a priori* specified elements. For a more general discussion of restricted and unrestricted solutions, see Jöreskog (1969).

Solving (3.15) for ξ and substituting this into (3.16) gives

$$x_2 = \Lambda_2 x_1 + u , \qquad (3.17)$$

where $u = \delta_2 - \Lambda_2 \delta_1$. Each equation in (3.17) is of the form (3.2) but it is not a regression equation because u is correlated with x_1, since δ_1 is correlated with x_1.

3.2 EXPLORATORY FACTOR ANALYSIS

Hägglund (1982) showed that instrumental variables can be obtained as follows. Let

$$x_i = \lambda'_i \mathbf{x}_1 + u_i , \qquad (3.18)$$

be the i-th equation in (3.17), where λ'_i is the i-th row of Λ_2, and let $\mathbf{x}_{(i)}(q - 1 \times 1)$ be a vector of the remaining variables in $\mathbf{x}_2$. Then u_i is uncorrelated with $\mathbf{x}_{(i)}$ so that $\mathbf{x}_{(i)}$ can be used as instrumental variables for estimating (3.18). Provided $q \geq k + 1$, this can be done for each $i = 1, 2, \ldots, q$.

The factors in a reference variables solution are neither standardized nor uncorrelated. After Λ has been estimated, the covariance matrix Φ and the error covariance matrix Θ can be estimated directly (non-iteratively) by unweighted or generalized least squares because the covariance structure in (3.14) is linear in Φ and Θ, see Browne (1974). Since most people prefer to interpret factors that are standardized, the solution is rescaled to standardized factors in the output.

The TSLS solution described here is used in LISREL to obtain starting values for the iterative methods, *i.e.*, the free elements of Λ_y and Λ_x in the model are estimated by the method descibed here and the elements of $\mathbf{B}$, Γ, Φ, and Ψ are estimated by the method described in Section 3.1. Note that this requires that $p \geq m + 1$ and $q \geq n + 1$, using the notation on p. 2 in Jöreskog & Sörbom (1996b). Whenever these conditions are not satisfied other methods are used to produce starting values.

Example: Exploratory Factor Analysis of Nine Psychological Variables

To illustrate exploratory factor analysis we use a classical data set. Holzinger & Swineford (1939) collected data on twenty-six psychological tests administered to 145 seventh- and eighth-grade children in the Grant-White school in Chicago. Nine of these tests are selected for this example. The nine selected variables and their intercorrelations are given in Table 3.4.[6]

---

[6]The correlations given here differ slightly from those given in Table 1.5 in Jöreskog & Sörbom (1996c) and Table 3.7 in Jöreskog & Sörbom (1996b), because the former have

Table 3.4 Correlation Matrix for Nine Psychological Variables

VIS PERC	1.000								
CUBES	0.326	1.000							
LOZENGES	0.449	0.417	1.000						
PAR COMP	0.342	0.228	0.328	1.000					
SEN COMP	0.309	0.159	0.287	0.719	1.000				
WORDMEAN	0.317	0.195	0.347	0.714	0.685	1.000			
ADDITION	0.104	0.066	0.075	0.209	0.254	0.178	1.000		
COUNTDOT	0.308	0.168	0.239	0.104	0.198	0.121	0.587	1.000	
S-C CAPS	0.487	0.248	0.373	0.314	0.356	0.222	0.418	0.528	1.000

In a completely exploratory factor analysis, both the number of factors and the reference variables are unknown and must be determined from the data. But to begin with, we shall assume that both of these are known and later show how to handle the completely exploratory case.

We assume that there are three factors and we use VIS PERC, PAR COMP, and ADDITION as reference variables. Thus all factor loadings for these variables are known. To estimate the factor loadings for the other variables, we estimate the relationship between each of these and the reference variables using all the others as instrumental variables. This can be done with the following RG commands in LISREL, where the index numbers of the variables are used instead of the names of the variables:

```
RG 2 on 1 4 7 with 3 5 6 8 9
RG 3 on 1 4 7 with 2 5 6 8 9
RG 5 on 1 4 7 with 2 3 6 8 9
RG 6 on 1 4 7 with 2 3 5 8 9
RG 8 on 1 4 7 with 2 3 5 6 9
RG 9 on 1 4 7 with 2 3 5 6 8
```

If there are many variables and factors this is not practical. We have therefore made it completely automatic. In the SIMPLIS command language, simply write

been computed from the raw scores (see file NPV1.PR2) whereas those in the previous tables have been copied from published correlation tables. There is no doubt that the correlations given here are the correct ones.

3.2 EXPLORATORY FACTOR ANALYSIS

```
Factor Analysis with 3 Factors
```

In the LISREL command language and in PRELIS, write

```
FA NF=3
```

LISREL will determine a suitable set of reference variables by a promax rotation of the maximum likelihood solution. Some users may prefer to use the unrotated solution, the varimax solution, or the promax solution, so all four solutions are given in the output.

Suppose, the correlation matrix in Table 3.4 is stored in the file NPV.KM. Then a SIMPLIS command file for factor analyzing the variables in this table is (see file NPV2.SPL):

```
Exploratory Factor Analysis of Nine Psychological Variables
Observed Variables
  'VIS PERC' CUBES    LOZENGES 'PAR COMP' 'SEN COMP'
  WORDMEAN   ADDITION COUNTDOT 'S-C CAPS'
Correlation Matrix from File NPV.KM
Sample Size 145
Factor Analysis with 3 Factors
End of Problem
```

The results are:

```
Maximum Likelihood Factor Analysis for 3 Factors

Unrotated Factor Loadings
```

	Factor 1	Factor 2	Factor 3	Unique Var
VIS PERC	0.52	0.16	0.45	0.50
CUBES	0.33	0.08	0.38	0.74
LOZENGES	0.49	0.08	0.47	0.54
PAR COMP	0.81	-0.32	-0.07	0.24
SEN COMP	0.79	-0.21	-0.15	0.30
WORDMEAN	0.76	-0.30	-0.06	0.32
ADDITION	0.42	0.54	-0.38	0.39
COUNTDOT	0.42	0.71	-0.06	0.32
S-C CAPS	0.57	0.43	0.16	0.46

```
Minimum Fit Function Chi-Square with 12 Degrees of Freedom =     9.31
```

Varimax-Rotated Factor Loadings

	Factor 1	Factor 2	Factor 3	Unique Var
VIS PERC	0.20	0.16	0.66	0.50
CUBES	0.11	0.05	0.50	0.74
LOZENGES	0.21	0.08	0.64	0.54
PAR COMP	0.84	0.07	0.23	0.24
SEN COMP	0.79	0.19	0.18	0.30
WORDMEAN	0.79	0.07	0.23	0.32
ADDITION	0.18	0.76	-0.04	0.39
COUNTDOT	0.00	0.78	0.27	0.32
S-C CAPS	0.19	0.53	0.48	0.46

Promax-Rotated Factor Loadings

	Factor 1	Factor 2	Factor 3	Unique Var
VIS PERC	0.68	0.04	0.03	0.50
CUBES	0.53	-0.01	-0.05	0.74
LOZENGES	0.67	0.06	-0.06	0.54
PAR COMP	0.07	0.85	-0.04	0.24
SEN COMP	0.01	0.81	0.09	0.30
WORDMEAN	0.08	0.80	-0.05	0.32
ADDITION	-0.18	0.13	0.79	0.39
COUNTDOT	0.20	-0.14	0.77	0.32
S-C CAPS	0.43	0.04	0.45	0.46

Factor Correlations

	Factor 1	Factor 2	Factor 3
Factor 1	1.00		
Factor 2	0.44	1.00	
Factor 3	0.34	0.26	1.00

Reference Variables Factor Loadings Estimated by TSLS

	Factor 1	Factor 2	Factor 3	Unique Var
VIS PERC	0.68	0.00	0.00	0.53
CUBES	0.60	-0.07	-0.18	0.72
	(0.25)	(0.16)	(0.19)	
	2.43	-0.44	-0.94	
LOZENGES	0.69	0.05	-0.14	0.54
	(0.28)	(0.16)	(0.20)	
	2.43	0.30	-0.69	
PAR COMP	0.00	0.87	0.00	0.24

3.2 EXPLORATORY FACTOR ANALYSIS

SEN COMP	-0.07	0.82	0.18	0.29
	(0.17)	(0.12)	(0.13)	
	-0.41	6.88	1.43	
WORDMEAN	-0.01	0.83	0.01	0.32
	(0.16)	(0.11)	(0.12)	
	-0.05	7.41	0.10	
ADDITION	0.00	0.00	0.78	0.40
COUNTDOT	0.39	-0.23	0.62	0.39
	(0.20)	(0.12)	(0.23)	
	2.01	-1.93	2.66	
S-C CAPS	0.51	-0.04	0.40	0.45
	(0.21)	(0.13)	(0.14)	
	2.47	-0.32	2.82	

Factor Correlations

	Factor 1	Factor 2	Factor 3
Factor 1	1.00		
Factor 2	0.57	1.00	
Factor 3	0.38	0.23	1.00

The first solution is the unrotated solution computed using the maximum likelihood procedure described by Jöreskog (1967) and in more detail by Jöreskog (1977). The second solution is the varimax solution of Kaiser (1958). Both of these are orthogonal solutions, *i.e.*, the factors are uncorrelated. The third solution is the promax solution of Hendrickson & White (1964). This is an oblique solution, *i.e.*, the factors are correlated. The varimax and the promax solutions are transformations of the unrotated solution and as such they are still maximum likelihood solutions. The fourth solution is the TSLS solution obtained in reference variables form as described earlier. The reference variables are chosen as those variables in the promax solution that have the largest factor loadings in each column. This gives VIS PERC, PAR COMP, and ADDITION as reference variables. The advantage of the TSLS solution is that standard errors can be obtained for all the variables except for the reference variables. This makes it easy to determine which loadings are statistically significant or not. The standard errors are given in parentheses below the loading estimate and the t-values are given below the standard errors. A simple rule to follow is to judge a factor loading statistically significant if its t-value is larger than 2 in magnitude. On the basis of the TSLS solution one can formulate an hypothesis for confirmatory factor analysis by specifying that all non-significant loadings are zero. This hypothesis should be tested on an independent sample.

The same result will be obtained with the following LISREL command file (see file NPV3.LS8):

```
Factor Analysis of Nine Psychological Variables
DA NI=9 NO=145
LA; 'VIS PERC' CUBES     LOZENGES 'PAR COMP'
    'SEN COMP' WORDMEAN ADDITION COUNTDOT 'S-C CAPS'
KM=NPV.KM;  FA NF=3
OU
```

The reference variables solution given in the output is a TSLS solution. It is not a maximum likelihood solution. For exploratory factor analysis, the TSLS solution is often quite sufficient. However, one can obtain the maximum likelihood solution for the reference variables representation using the SIMPLIS command file NPV4.SPL or the LISREL command file NPV5.LS8. These files are not shown here. Because the correlation matrix is analyzed in these examples, the standard errors for the factor loadings are slightly incorrect, see Cudeck (1989). Correct standard errors can be obtained, still using the correlation matrix, by the more complicated LISREL command file NPV6.LS8. Alternatively, the problem of incorrect standard errors can be avoided entirely by analyzing the covariance matrix instead of the correlation matrix. To obtain standardized factor loadings put SC on the OU command in LISREL or on an Options line in SIMPLIS.

If the correlation matrix has not been computed but raw data is available, one can use the following PRELIS command file (see NPV7A.PR2) to obtain the same result directly (here it is assumed that the raw data is in the file NPV.RAW):

```
Factor Analysis of Nine Psychological Variables
DA NI=9
LA; 'VIS PERC' CUBES     LOZENGES 'PAR COMP' 'SEN COMP'
       WORDMEAN ADDITION COUNTDOT 'S-C CAPS'
RA=NPV.RAW
CO ALL; FA NF=3
OU MA=KM KM=NPV.KM
```

This will also save the computed correlation matrix in a file so that if one wants to reanalyze the data with a different number of factors one can do this with LISREL rather than PRELIS. Regardless of whether PRELIS or LISREL is used, one can read in a large number of variables and use an SE command to select a subset of variables to factor analyze.[7]

3.2.1 Number of Factors

There is no unique way to determine the number of factors. This is best done by the investigator who knows what the variables (are supposed to) measure. Then the number of factors can be specified *a priori* at least tentatively. Many procedures have been suggested in the literature to determine the number of factors analytically. One of them is to continue to extract factors until there no longer are at least three large loadings in each column of the varimax solution, say. The question is what is meant by large. The TSLS reference variables solution offers an answer to this question by considering as large those loadings which are statistically significant. Our procedure for deciding on the number of factors is based on statistical fit.

If the number of factors is not specified, *i.e.*, if with 3 Factors or NF=3 is omitted, PRELIS or LISREL will try to determine a suitable number of factors using a decision procedure based on a number of fit criteria for maximum likelihood factor analysis for $k = 0, 1, \ldots, k_{max}$, where k_{max} is the largest number of factors for which a factor solution can be obtained. Note that only the solution for the number of factors determined in this way will appear in the output. If one wants a solution with a larger or smaller number of factors than that determined by this procedure, one must redo the analysis and specify the number of factors.

For our NPV example (see file NPV7B.PR2), the decision table for deciding the number of factors is based on the values in Table 3.5.

The quantities c_k, d_k, P_k, Δc_k, Δd_k, $P_{\Delta c}$, and ρ_k, are defined as follows.

---

[7]The SE command is new in PRELIS 2.30 and has the same syntax as the SE command in LISREL

$$c_k = [n - (2p+5)/6 - 2k/3][\ln|\hat{\Sigma}| - \ln|\mathbf{S}|], \quad (3.19)$$
$$k = 0, 1, \ldots, k_{max}$$
$$d_k = [(p-k)^2 - (p+k)]/2, \, k = 0, 1, \ldots, k_{max} \quad (3.20)$$
$$P_k = Pr\{\chi^2_{d_k} > c_k\}, \, k = 0, 1, \ldots, k_{max} \quad (3.21)$$
$$\Delta c_k = c_k - c_{k-1}, \, k = 1, 2, \ldots, k_{max} \quad (3.22)$$
$$\Delta d_k = d_k - d_{k-1}, \, k = 1, 2, \ldots, k_{max} \quad (3.23)$$
$$P_{\Delta c} = Pr\{\chi^2_{\Delta d_k} > \Delta c_k\}, \, k = 1, 2, \ldots, k_{max} \quad (3.24)$$
$$\rho_k = \sqrt{[(c_k - d_k)/nd_k]}, \, k = 0, 1, \ldots, k_{max} \quad (3.25)$$

Table 3.5   Fit Statistics for Deciding the Number of Factors

k	c_k	d_k	P_k	Δc_k	Δd_k	$P_{\Delta c}$	ρ_k
0	488.91	36	0.000				0.295
1	175.49	27	0.000	313.41	9	0.000	0.195
2	61.70	19	0.000	113.79	8	0.000	0.124
3	9.38	12	0.670	52.32	7	0.000	0.000
4	2.59	6	0.858	6.79	6	0.341	0.000

Here c_k is the chi-square statistic for testing the fit of k factors, see Lawley & Maxwell (1971, pp. 35–36). If the model holds and the variables have a multivariate normal distribution, this is distributed in large samples as χ^2 with d_k degrees of freedom.[8] The P-value of this test is P_k, i.e., the probability that a random χ^2 with d_k degrees of freedom exceeds the chi-square value actually obtained. For reasons stated elsewhere (see, e.g., Jöreskog & Sörbom, 1996b, p. 28, or Jöreskog & Sörbom, 1996c, p. 122), it is better to regard these quantities as approximate measures of fit rather than as test statistics. Δc_k measures how much better the fit is with k factors than with $k-1$ factors. Δd_k and $P_{\Delta c}$ are the corresponding degrees of freedom and P-value. ρ_k is Steiger's (1990) *Root Mean Squared Error of Approximation* (RMSEA) which is a measure of population error

---

[8]For $k = 0$ this is a test of the hypothesis that the variables are uncorrelated. If this hypothesis cannot be rejected, it is meaningless to do a factor analysis.

3.2 EXPLORATORY FACTOR ANALYSIS

per degree of freedom, see Browne & Cudeck (1993) or Jöreskog & Sörbom (1996c).

LISREL investigates these quantities for $k = 1, 2, \ldots, k_{max}$ and determines the smallest acceptable k with the following decision procedure: If $P_k > .10$, k factors are accepted. Otherwise, if $P_{\Delta c} > .10$, $k-1$ factors are accepted. Otherwise, if $\rho_k < .05$, k factors are accepted. If none of these conditions are satisfied, k is increased by 1.

The first criterion, $P_k > .10$, guarantees that one stops at k if the overall fit is good. The second criterion, $P_{\Delta c} > .10$, guarantees that one will not increase the number of factors unless the improvement in fit is statistically significant at the 10% level. The third criterion, $\rho_k < .05$, is the Browne–Cudeck guideline (Browne & Cudeck, 1993, p. 144). This guarantees that one does not get too many factors in large samples. This procedure may not give a satisfactory answer to the number of factors in all respects, but at least there will not be a tendency to overfit, *i.e.*, to take too many factors.

For the values in Table 3.5 the decision will be $k = 3$ factors, because for $k = 2$, P_k and $P_{\Delta c}$ are too small and ρ_k is too large, but for $k = 3$, P_k is acceptable.

In the output (see file NPV7B.OUT), the decision table is given as:

```
Decision Table for Number of Factors

Factors    Chi2    df      P       DChi2  Ddf    PD      RMSEA
-------    ----    --      -       -----  ---    --      -----
   0      488.91   36    0.000                            0.295
   1      175.49   27    0.000    313.41   9    0.000    0.195
   2       61.70   19    0.000    113.79   8    0.000    0.124
   3        9.38   12    0.670     52.32   7    0.000    0.000
   4        2.59    6    0.858      6.79   6    0.341    0.000
```

3.2.2 Factor Scores

To obtain factor scores for the factors in the TSLS reference variables solution, add the keyword FS on the FA command in PRELIS. Factor scores can only be obtained by PRELIS because it requires raw data, but see Section 3.6 on how to obtain sample scores for all the latent variables in any LISREL model.

The factor scores are computed by an extension of a formula given by Anderson & Rubin (1956). These factor scores are unbiased estimates of the factors and their sample covariance matrix is exactly equal to the estimated covariance or correlation matrix of the reference variables factors. The other two commonly used methods for estimating factor scores, *i.e.*, the regression method and Bartlett's method, do not have these properties.

To obtain factor scores just add FS on the FA command in files NPV7A.PR2 or NPV7B.PR2, see file NPV7C.PR2. The factor scores are saved as a plain text (ASCII) file with the same name as the input file, but with suffix FSC. Thus, in this case the factor scores will be saved in the file NPV7C.FSC. This can be read in free format.

Possible uses of these factor scores are

- Select subgroups of individuals on the basis of the factor scores
- Rank the individuals on the basis of the factor scores for one factor
- Correlate the factor scores with some external variable
- Compute scores for the unique (error) variables δ

Here we illustrate how to merge the factor scores with the scores on the observed variables and compute the joint correlation matrix of the observed variables and the factor scores. The PRELIS command file to do this is (see file NPV7D.PR2):

```
Merging Observed Variables and Factor Scores and
Computing Joint Correlation Matrix
Data NI = 9,3
Labels
'VIS PERC' CUBES LOZENGES 'PAR COMP' 'SEN COMP'
WORDMEAN  ADDITION COUNTDOT 'S-C CAPS'
Factor_1 Factor_2 Factor_3
Rawdata=NPV.RAW,NPV7C.FSC
Continuous 'VIS PERC' - Factor_3
Output MA=KM
```

Note that the correlations among the three factors are exactly the same as the correlation matrix for the factors of the TSLS reference variables solution, see any of the files NPV2.OUT – NPV6.OUT.

3.3 Principal Components

Principal components are uncorrelated linear combinations of observed variables that account for maximum variance in the observed variables. For further explanation of principal components and their interpretation, we refer to Mardia, Kent, & Bibby (1980, Chapter 8), Reyment & Jöreskog (1993), or any other book covering principal components. For an explanation of the difference between principal components and factor analysis, see Jöreskog (1979).

Unlike factor analysis, which is scale independent in a certain sense, the results of principal components depend on the unit of measurement in the observed variables. For example, the analysis of the covariance matrix and the correlation matrix can give very different results, especially if the variances differ much across variables. This is the case in our example. Principal components are best applied to variables which are measured in the same units of measurements and which have approximately the same variances. In other situations, it is best to use the correlation matrix. Principal components are sometimes applied to the moment matrix instead of the covariance or correlation matrix, *i.e.*, the means are not subtracted from the observed scores before analysis. This too can give a different result compared with the covariance or correlation matrix.

Principal components may be obtained by including a PC command in a PRELIS, SIMPLIS, or LISREL command file. The syntax of the PC command is similar to that of the FA command described in the previous section. In PRELIS the syntax is

PC NC=k PS

where NC=k is used to specify that the first k principal components are to be computed. If NC=k is omitted, all principal components are computed, *i.e.*, $k = p$, where p is the number of observed variables. With PS on the PC command, the scores of the principal components are computed for each unit in the sample and saved in the file input.PSC, where input is the name of the PRELIS command file (without the suffix). This file can be read by PRELIS in free format.

In the LISREL command language the syntax is the same, except that PS is not allowed since individual component scores can only be computed from raw data.

In the SIMPLIS command language the syntax is

```
Principal Components with k Components
```

Example: Principal Components of Five Meteorological Variables

Mardia, Kent, & Bibby (1980, p. 248) give the data on five meteorological variables presented in Table 3.6. The corresponding unbiased covariance matrix is given in Table 3.7. The variables are

$x_1 =$ rainfall in November and December (in millimeters)

$x_2 =$ avarage July temperature (in degrees Celsius)

$x_3 =$ rainfall in July (in millimeters)

$x_4 =$ radiation in July (in millimeters of alcohol)

$x_5 =$ average harvest yield (in quintals per hectare)

Table 3.6   Raw Data in Five Meteorological Variables

Year	x_1	x_2	x_3	x_4	x_5
1920–21	87.9	19.6	1.0	1661	28.37
1921–22	89.9	15.2	90.1	968	23.77
1922–23	153.0	19.7	56.6	1353	26.04
1923–24	132.1	17.0	91.0	1293	25.74
1924–25	88.8	18.3	93.7	1153	26.68
1925–26	220.9	17.8	106.9	1286	24.29
1926–27	117.7	17.8	65.5	1104	28.00
1927–28	109.0	18.3	41.8	1574	28.37
1928–29	156.1	17.8	57.4	1222	24.96
1929–30	181.5	16.8	140.6	902	21.66
1930–31	181.4	17.0	74.3	1150	24.37

We illustrate how the principal components can be obtained for the covariance matrix using a LISREL command file (see file PCEX1.LS8):

3.3 PRINCIPAL COMPONENTS

Table 3.7   Covariance Matrix of Five Meteorological Variables

x_1	x_2	x_3	x_4	x_5
1973.298				
-4.921	1.637			
799.564	-29.279	1346.859		
-2439.351	217.198	-6822.728	52914.656	
-57.214	1.735	-62.080	361.803	4.496

```
Example of Principal Components
DA NI=5 NO=11
CM
   1973.298
     -4.921    1.637
    799.564  -29.279   1346.859
  -2439.351  217.198  -6822.728  52914.656
    -57.214    1.735    -62.080    361.803   4.496
PC
OU
```

The results are given in the output as follows.

```
Principal Component Analysis

Eigenvalues and Eigenvectors

              PC_1       PC_2      PC_3      PC_4      PC_5
           --------   --------  --------  --------  --------
Eigenvalue 53927.94    1999.96    311.26      1.26      0.52
% Variance    95.89       3.56      0.55      0.00      0.00
Cum. % Var    95.89      99.44    100.00    100.00    100.00
           --------   --------  --------  --------  --------
   VAR 1      -0.05       0.95     -0.30      0.01     -0.01
   VAR 2       0.00       0.00     -0.01      0.51      0.86
   VAR 3      -0.13       0.29      0.95      0.02      0.00
   VAR 4       0.99       0.08      0.11     -0.01      0.00
   VAR 5       0.01      -0.02     -0.01      0.86     -0.51
```

Correlations between Variables and Principal Components

	PC_1	PC_2	PC_3	PC_4	PC_5
VAR 1	-0.25	0.96	-0.12	0.00	0.00
VAR 2	0.74	0.09	-0.11	0.45	0.49
VAR 3	-0.82	0.35	0.46	0.00	0.00
VAR 4	1.00	0.02	0.01	0.00	0.00
VAR 5	0.75	-0.44	-0.06	0.46	-0.17

Variance Contributions

	PC_1	PC_2	PC_3	PC_4	PC_5
VAR 1	0.06	0.92	0.01	0.00	0.00
VAR 2	0.54	0.01	0.01	0.20	0.24
VAR 3	0.67	0.12	0.21	0.00	0.00
VAR 4	1.00	0.00	0.00	0.00	0.00
VAR 5	0.56	0.20	0.00	0.21	0.03

The first line gives the eigenvalues of the covariance matrix. These are the variances of the principal components. The second line gives the eigenvalues in percentage of the total variance and the third line gives these percentages cumulatively. The next five columns give the eigenvectors of the covariance matrix normalized so that their sum of squares is one and such that the largest value is positive. These eigenvectors are the coefficients (weights) in the linear combinations defining the principal components.

The next set of five columns shows the correlations between the observed variables and the principal components and the last set of five columns gives the variance contribution of each principal component to the variance of each observed variable. If all principal components are computed, as is the case here, these variance contributions sum to one row-wise.

The first two principal components account for 99.44% of the variance. The first component is dominated by the radiation variable x_4. The second component is a rainfall variable dominated by x_1 and x_3.

To analyze the correlation matrix instead of the covariance matrix of the five meteorological variables, just add MA=KM on the DA command in PCEX1.LS8 (see file PCEX2.LS8). We leave it to the reader to verify that this gives a very different result and interpretation.

One can also compute the component scores, *i.e.*, the scores on the linear combinations. For this, one must use PRELIS and the raw data. File

PCEX3.PR2 illustrates how to obtain the the component scores of the first two principal components:

```
Principal Components with Components Scores
DA NI=5
RA=PCEX.RAW
CO ALL
PC NC=2 PS
OU MA=CM
```

The component scores are given in the file PCEX3.PSC:

```
  423.256    -6.692
 -275.908   -12.942
  111.016     8.089
   49.359    20.667
  -86.970   -38.177
   38.374    66.545
 -131.071  -101.073
  335.221    -4.458
  -20.796     2.797
 -348.384    55.395
  -94.095     9.848
```

3.4 Normal Scores

The analysis of continuous non-normal variables in structural equation models is problematic in several ways. If the maximum likelihood (ML) method is used, standard errors and chi-squares may be incorrect. In theory, weighted least squares (WLS or ADF) with a correct weight matrix should produce correct estimates of standard errors and chi-squares, but this requires a very large sample. Sometimes a reasonable compromise is to use ML despite the non-normality and correct for the bias in standard errors, but this too requires a very large sample.

Another solution to non-normality is to normalize the variables before analysis. Normal scores offer an effective way of normalizing a continuous variable for which the origin and unit of measurement have no intrinsic meaning, such as test scores. This section illustrates how this works.

Consider a data matrix of N cases on p variable after listwise deletion of missing values, if any. To begin with, we assume that the variables are continuous variables. Consider anyone of these p variables to be normalized, and let

$$x_1, x_2, \ldots, x_N$$

be the sample values. Suppose there are k distinct values

$$x_1, x_2, \ldots, x_k$$

in ascending order, and let n_i be the frequency of occurence of x_i, i.e., the number of times the value x_i occurs in the sample. Each $n_i \geq 1$ and $\sum_{i=1}^{k} n_i = N$. The *normal score* $\hat{z}_i$ corresponding to x_i is computed by the formula

$$\hat{z}_i = (N/n_i)[\phi(\hat{\alpha}_{i-1}) - \phi(\hat{\alpha}_i)] \qquad i = 1, 2, \ldots, k, \qquad (3.26)$$

where $\hat{\alpha}_0 = -\infty$, $\hat{\alpha}_k = +\infty$, and

$$\hat{\alpha}_i = \Phi^{-1}\left(\sum_{j=1}^{i} n_j/N\right) \qquad i = 1, 2, \ldots, k-1. \qquad (3.27)$$

Here ϕ is the standard normal density function and Φ^{-1} is the inverse standard normal distribution function. PRELIS scales the normal scores so that they have the same sample mean and standard deviation as the original variable. Thus, the normal score is a monotonic transformation of the original score with the same mean and standard deviation but with much reduced skewness and kurtosis. The rank ordering of cases are the same.

The following example illustrates how it works.

Example: Normalizing the Nine Psychological Variables

To normalize the raw scores of the nine psychological variables in file NPV.RAW, use the following PRELIS command file (see NPVNSC1.PR2):

3.4 NORMAL SCORES

```
Compute Correlation Matrix of the Normalized Scores
for Nine Psychological Variables
Data Ni = 9
Labels
'VIS PERC' CUBES LOZENGES 'PAR COMP' 'SEN COMP'
WORDMEAN  ADDITION COUNTDOT 'S-C CAPS'
Rawdata=NPV.RAW
Continuous 'VIS PERC' - 'S-C CAPS'
NScores 'VIS PERC' - 'S-C CAPS'
Output MA=KM
```

This run will normalize all the variables in NPV.RAW and compute the sample correlation matrix of the normalized variables. The normalized variables will have the same name as the observed variables (see next example on how to save both the original variables and their normal scores). One can also save the normal scores for every case in the data by putting RA=*filename* on the Output line.

The general syntax for the NS command is

NS *varlist*

where *varlist* is a list of variables, *i.e.*, any subset of the variables that has been read or constructed with NE commands.

To see the effects of the normalization, we list here the characteristics of the variables before and after the normalization. From the output of NPV1.PR2, we obtain these characteristics of the variables before normalization:

Univariate Summary Statistics for Continuous Variables

Variable	Mean	St. Dev.	T-Value	Skewness	Kurtosis
VIS PERC	29.579	6.914	51.517	-0.119	-0.046
CUBES	24.800	4.445	67.183	0.239	0.872
LOZENGES	15.966	8.317	23.115	0.623	-0.454
PAR COMP	9.952	3.375	35.502	0.405	0.252
SEN COMP	18.848	4.649	48.817	-0.550	0.221
WORDMEAN	17.283	7.947	26.186	0.729	0.233
ADDITION	90.179	23.782	45.660	0.163	-0.356
COUNTDOT	109.766	20.995	62.955	0.698	2.283
S-C CAPS	191.779	37.035	62.355	0.200	0.515

Test of Univariate Normality for Continuous Variables

	Skewness		Kurtosis		Skewness and Kurtosis	
Variable	Z-Score	P-Value	Z-Score	P-Value	Chi-Square	P-Value
VIS PERC	-0.604	0.546	0.045	0.964	0.367	0.833
CUBES	1.202	0.229	1.843	0.065	4.842	0.089
LOZENGES	2.958	0.003	-1.320	0.187	10.491	0.005
PAR COMP	1.995	0.046	0.761	0.447	4.559	0.102
SEN COMP	-2.646	0.008	0.693	0.489	7.483	0.024
WORDMEAN	3.385	0.001	0.720	0.472	11.977	0.003
ADDITION	0.826	0.409	-0.937	0.349	1.560	0.458
COUNTDOT	3.263	0.001	3.325	0.001	21.699	0.000
S-C CAPS	1.008	0.313	1.273	0.203	2.638	0.267

Test of Multivariate Normality for Continuous Variables

Skewness			Kurtosis			Skewness and Kurtosis	
Value	Z-Score	P-Value	Value	Z-Score	P-Value	Chi-Square	P-Value
11.733	5.426	0.000	106.098	3.023	0.003	38.579	0.000

Before normalization, several of the variables have significant skewness and kurtosis. In particular, WORDMEAN has the highest skewness and COUNTDOT has both large skewness and large kurtosis. As a consequence, all tests of multivariate normality are strongly rejected.

From the output of NPVNSC1.PR2 we get these characteristics of the variables after normalization:

Univariate Summary Statistics for Continuous Variables

Variable	Mean	St. Dev.	T-Value	Skewness	Kurtosis
VIS PERC	29.579	6.914	51.517	0.004	-0.037
CUBES	24.800	4.445	67.183	-0.002	-0.064
LOZENGES	15.966	8.317	23.115	0.011	-0.062
PAR COMP	9.952	3.375	35.502	0.003	-0.018
SEN COMP	18.848	4.649	48.817	-0.008	-0.027
WORDMEAN	17.283	7.947	26.186	0.005	-0.028
ADDITION	90.179	23.782	45.660	0.000	-0.022
COUNTDOT	109.766	20.995	62.955	0.001	-0.023
S-C CAPS	191.779	37.035	62.355	0.000	-0.022

3.4 NORMAL SCORES

Test of Univariate Normality for Continuous Variables

	Skewness		Kurtosis		Skewness and Kurtosis	
Variable	Z-Score	P-Value	Z-Score	P-Value	Chi-Square	P-Value
VIS PERC	0.019	0.985	0.067	0.946	0.005	0.998
CUBES	-0.011	0.991	-0.005	0.996	0.000	1.000
LOZENGES	0.055	0.956	0.000	1.000	0.003	0.998
PAR COMP	0.016	0.988	0.119	0.905	0.014	0.993
SEN COMP	-0.038	0.969	0.096	0.924	0.011	0.995
WORDMEAN	0.025	0.980	0.092	0.927	0.009	0.995
ADDITION	0.000	1.000	0.108	0.914	0.012	0.994
COUNTDOT	0.003	0.998	0.106	0.916	0.011	0.994
S-C CAPS	0.000	1.000	0.108	0.914	0.012	0.994

Test of Multivariate Normality for Continuous Variables

Skewness			Kurtosis			Skewness and Kurtosis	
Value	Z-Score	P-Value	Value	Z-Score	P-Value	Chi-Square	P-Value
8.187	1.737	0.082	100.807	1.318	0.188	4.754	0.093

It is seen that after normalization none of the variables have any essential skewness or kurtosis and the assumption of multivariate normality cannot be rejected. Note also that the means and standard deviations of the normalized scores are exactly equal to those of the original variables. The correlation matrix of the normalized variables is given in Table 3.8. These correlations should be compared with those of the original variables in Table 3.4. It is seen that the correlations are similar but not equal. This shows that the normal scores are not linear functions of the original scores but nearly so. We will investigate this relationship in the next example. We leave it to the reader to do a factor analysis of the correlations in Table 3.8 with TSLS and ML as in files NPV2.SPL and NPV4.SPL and compare the results.

Example: Normalizing WORDMEAN and COUNTDOT

We now take a closer look at the normalization of the two most non-normal variables, WORDMEAN and COUNTDOT. Recall that WORDMEAN has significant skewness and COUNTDOT has significant skewness and kurtosis. This example will save the raw data on the normalized variables as well

Table 3.8
Correlation Matrix for Normalized Nine Psychological Variables

VIS PERC	1.000								
CUBES	0.333	1.000							
LOZENGES	0.455	0.403	1.000						
PAR COMP	0.337	0.240	0.321	1.000					
SEN COMP	0.311	0.175	0.300	0.720	1.000				
WORDMEAN	0.307	0.218	0.359	0.691	0.700	1.000			
ADDITION	0.103	0.081	0.088	0.195	0.264	0.201	1.000		
COUNTDOT	0.291	0.188	0.250	0.116	0.207	0.166	0.603	1.000	
S-C CAPS	0.489	0.248	0.381	0.300	0.345	0.257	0.429	0.548	1.000

as the original variables so that all four variables can be studied simultaneously. This can be done with the following PRELIS command file (see file WMCD1.PR2):

```
Compute Normal Scores for WORDMEAN and COUNTDOT
Data NI = 9
Labels; 'VIS PERC' CUBES    LOZENGES 'PAR COMP' 'SEN COMP'
         WORDMEAN   ADDITION COUNTDOT 'S-C CAPS'
Rawdata=NPV.RAW;  Continuous All
New NSC(WM) = WORDMEAN            ! This line saves a copy of WORDMEAN
New NSC(CD) = COUNTDOT            ! This line saves a copy of COUNTDOT
NScores NSC(WM) NSC(CD)           ! This line normalizes the copies
Select WORDMEAN COUNTDOT NSC(WM) NSC(CD)
! In the end we have both the original variables and their normalized versions
Output MA=KM RA=WMCD.RAW
```

The output gives the correlation matrix of the four variables as:

```
Correlation Matrix

              WORDMEAN    COUNTDOT     NSC(WM)     NSC(CD)
              --------    --------    --------    --------
  WORDMEAN       1.000
  COUNTDOT       0.121       1.000
   NSC(WM)       0.981       0.122       1.000
   NSC(CD)       0.153       0.982       0.166       1.000
```

3.4 NORMAL SCORES

The correlation between the originals and their corresponding normal scores is 0.98 for both variables. The correlation between the two originals is 0.12 and that between their normal scores is 0.17. We now investigate the non-linear relationship between each original and its corresponding normal score (see file WMSD2.PR2):

```
Estimate Regression of
    NSC(WM) on WORDMEAN, WORDMEAN**2 WORDMEAN**3
    NSC(CD) on COUNTDOT, COUNTDOT**2 COUNTDOT**3
Data Ni = 4
Labels
WORDMEAN  COUNTDOT NSC(WM) NSC(CD)
Rawdata=WMCD.RAW
Continuous All
New WM2=WORDMEAN**2
New WM3=WORDMEAN**3
New CD2=COUNTDOT**2
New CD3=COUNTDOT**3
RG NSC(WM) on WORDMEAN WM2 WM3
RG NSC(CD) on COUNTDOT CD2 CD3
Output
```

The output gives the following results

```
Estimated Equations

    NSC(WM) = - 9.309  + 2.469*WORDMEAN - 0.0623*WM2 + 0.000733*WM3
             (0.271)   (0.0472)           (0.00250)     (0.0000397)
             -34.400    52.298             -24.922        18.436

            + Error, R² = 0.997

Error Variance = 0.165

    NSC(CD) = 24.512 - 0.0386*COUNTDOT + 0.0121*CD2 - 0.0000422*CD3
             (8.428)   (0.214)           (0.00175)    (0.00000456)
              2.908    -0.180             6.942         -9.249

            + Error, R² = 0.993

Error Variance = 3.031
```

This shows that the relationship between the normal raw score and the corresponding original score can be almost exactly represented by a cubic. Note the high value of R^2. The nature of this relationship can also be seen by importing the data file WMCD.RAW into PRELIS and then looking at the bivariate graph (line plot)[9] of WORDMEAN on NSC(WM) or of COUNTDOT on NSC(CD).

Normal scores can also be computed for ordinal variables. Then there will be one normal score for each distinct value (category) of the ordinal variable. The idea would be to use these normal scores as a continuous variable instead of the raw scores that represent the different categories in the data, but it is doubtful that this would be any better than to use the original scores.[10] Our recommendation is to treat ordinal variables as ordinal (and not as continuous) and analyze them as described by Jöreskog (1990,1994) or by Jöreskog & Sörbom (1996a).

3.5 System Files

PRELIS and LISREL generate several system files through which they can communicate with each other. These system files are binary files and can therefore be read very fast. Some of these system files can be used directly by users. Here we present these system files and their uses.

3.5.1 The PRELIS System File

A PRELIS system file, or a PSF file for short is created when the user specifies RA=*filename*.PSF on the OU command.[11]

Its use is recommended especially now that the analysis features in PRELIS have been expanded. Once the user finds the dataset in good shape, after

---

[9]Only the Windows version of LISREL has such capabilities built in.

[10]Normal scores for ordinal variables were available in PRELIS 1 but they were abandoned in PRELIS 2 for this reason. Now they have been put back for anyone who needs them.

[11]It is automatically created in the Windows interface of the program each time raw data is imported into a spreadsheet. PRELIS can import such data in a variety of different formats from many different programs, such as ASCII (plain text), SAS, or SPSS system files. The PRELIS system files have the suffix PSF. This suffix is added to the filename that the user specifies the first time the data is imported into PRELIS. Once created, the PSF file can be opened in a spreadsheet by selecting it.

3.5 SYSTEM FILES

taking care of missing values, outliers, recoding, etc., the creation of a PSF file makes further analysis of the data with PRELIS easier and faster. For example, all the multilevel analyses in Chapter 2 were done using a PSF file.

3.5.2 The Data System File

A data system file, or a DSF file for short, is created each time PRELIS is run. Its name is the same as the PRELIS command file but with the suffix DSF. The DSF file contains all the information about the variables that LISREL needs to analyze the data, *i.e.*, names, sample size, means, standard deviations, covariance or correlation matrix, and the location of the asymptotic covariance matrix, if any. The DSF file can be read by LISREL directly instead of reading the names, sample size, means, standard deviations, covariance or correlation matrix, and the asymptotic covariance matrix, if any, by separate commands in a LISREL command file.

To read a DSF file in the SIMPLIS command language, write:

```
System file from File filename.DSF
```

This line replaces the following typical lines in a SIMPLIS command file (other variations are possible):

```
Observed Variables: A B C D E F
Means from File filename
Covariance Matrix from File filename
Asymptotic Covariance Matrix from File filename
Sample Size: 678
```

To read a DSF file in the LISREL command language, write:

```
SY=filename.DSF
```

This replaces the following typical lines in a LISREL command file (other variations are possible):

```
DA NI=k  NO=n
ME=filename
CM=filename
AC=filename
```

As the DSF file is a binary file, it can be read much faster than the (ASCII) data files. To make optimal use of this, consider the following strategy, assuming the data consists of many variables, possibly several hundreds, and a very large sample.

Use PRELIS to deal with all problems in the data, *i.e.*, missing data, variable transformation, recoding, definition of new variables, etc., and compute means and the covariance matrix or correlation matrix, say, and the asymptotic covariance matrix, if needed. Read the DSF file into LISREL, select the variables for analysis and specify the model. Several different sets of variables may be analyzed in this way, each one being based on a small subset of the variables in the DSF file. The point is that there is no need to go back to PRELIS to compute new summary statistics for each LISREL model. With the SIMPLIS command language, selection of variables is automatic in the sense that only the variables included in the model will be used.

The use of the DSF is especially important in simulations as these will go much faster.

3.5.3 The Model System File

A model system file, or MSF file for short, is created each time a LISREL command file is run. Its name is the same as the LISREL command file but with the suffix MSF. The MSF file contains all the information about the model that LISREL needs to produce a path diagram, *i.e.*, type and form of each parameter, parameter estimates, standard errors, t-values, modification indices, fit statistics, etc. Users do not have a direct need for the MSF file.

3.6 Latent Variable Scores

Previously, we described how to obtain factor scores for the factors in exploratory factor analysis. In a similar way, one can obtain scores for the latent variables in any estimated LISREL model. To obtain the latent variable scores one must have both the raw scores and the estimated model. LISREL will automatically append the latent variable scores to the raw scores if the PSF file is specified in the LISREL or SIMPLIS command file. In the extended PSF file the η-variables appear first and then the ξ-variables. Care must be taken to specify the correct PSF file, *i.e.*, the one which is used to compute the covariance matrix on which the LISREL model is estimated. To specify the PSF file in a SIMPLIS syntax file, include the line

PSFfile *filename*.PSF

In LISREL syntax, include the line

PS= *filename*.PSF

The sample mean vector and covariance matrix of the latent variable scores estimated in this way are exactly equal to the mean vector and covariance matrix given in the output from LISREL run. One possible use of the latent variable scores is to estimate a structural equation model in two steps. First estimate the measurement model for the y- and x-variables and save the latent variables scores. Then the structural part of the model may be estimated using these scores as observed variables. An example of this follows in Section 3.7.2.

3.7 Interaction and Non-Linear Models

In recent years, there has been considerable interest in extending and applying structural equation models to situations where there are non-linear relationships involving latent variables, in particular to models with interaction effects. Various chapters in Schumacker & Marcoulides (1998) discuss the issues involved in non-linear structural equation models and describe different methods of estimation. The full information methods proposed by Jöreskog & Yang (1996) are difficult to apply in practice due to the complicated non-linear constraints that must be specified and the necessity to have large samples and use asymptotic covariance matrices. Bollen (1995, 1996) and Bollen & Paxton (1998) showed that non-linear models can be estimated easily with TSLS. Bollen & Paxton (1998) describe a rather complicated procedure to do this using SAS. Here we show how TSLS can be used easily and effectively using a single command in PRELIS 2.30.

To describe the basic idea of TSLS we use the Kenny–Judd model as a prototype for non-linear models (see Kenny & Judd, 1984). Bollen & Paxton (1998) demonstrated how TSLS can be used with several other types of non-linear models.

3.7.1 Estimation by TSLS

Example: The Kenny–Judd Model

The Kenny–Judd model is

$$y = \alpha + \gamma_1 \xi_1 + \gamma_2 \xi_2 + \gamma_3 \xi_1 \xi_2 + \zeta, \qquad (3.28)$$

where ξ_1 and ξ_2 are latent variables and ζ is a random error term assumed to be uncorrelated with ξ_1 and ξ_2. Kenny & Judd (1984) considered the case when there are two observable indicators x_1 and x_2 of ξ_1 and two observable indicators x_3 and x_4 of ξ_2, such that

$$\begin{pmatrix} x_1 \\ x_2 \\ x_3 \\ x_4 \end{pmatrix} = \begin{pmatrix} \tau_1 \\ \tau_2 \\ \tau_3 \\ \tau_4 \end{pmatrix} + \begin{pmatrix} 1 & 0 \\ \lambda_2 & 0 \\ 0 & 1 \\ 0 & \lambda_4 \end{pmatrix} \begin{pmatrix} \xi_1 \\ \xi_2 \end{pmatrix} + \begin{pmatrix} \delta_1 \\ \delta_2 \\ \delta_3 \\ \delta_4 \end{pmatrix}. \qquad (3.29)$$

Kenny & Judd (1984) did not include the constant intercept terms α and τ_i in (3.28) and (3.29), but as argued by Jöreskog & Yang (1996), these are necessary for a correct analysis of the model.

3.7 INTERACTION AND NON-LINEAR MODELS

Solving for ξ_1 and ξ_2 in the first and third equation in (3.29) and substituting into (3.28) gives

$$y_1 = \alpha + \gamma_1 x_1 + \gamma_2 x_3 + \gamma_3 x_1 x_3 + u, \qquad (3.30)$$

where

$$u = -\gamma_1 \delta_1 - \gamma_2 \delta_3 - \gamma_3 (x_1 \delta_3 + x_3 \delta_1 - \delta_1 \delta_3) + \zeta. \qquad (3.31)$$

Note that the γ's in (3.30) are the same as in (3.28) but the error term u is not the same as ζ. The error term u in (3.30) is correlated with the variables on the right side in (3.30) so that ordinary least squares cannot be used to estimate this equation. However, as shown by Bollen & Paxton (1998), x_2, x_4, and $x_2 x_4$ can be used as instrumental variables. If the model holds, these instrumental variables are uncorrelated with u. Assuming that raw data on y, x_1, x_2, x_3, and x_4, are available in the file KJUDD.RAW, the following PRELIS command file computes the necessary product variables and estimates the non-linear equation (3.30) directly (see file KJTSLS1.PR2):

```
Estimating Kenny-Judd Model by Bollen's TSLS
! Note the use of the semicolon as command delimiter
DA NI=5
LA; Y X1 X2 X3 X4
RA=KJUDD.RAW;    CO ALL
NE X1X3=X1*X3;  NE X1X4=X1*X4;  NE X2X3=X2*X3;  NE X2X4=X2*X4
RG Y ON X1 X3 X1X3 WITH X2 X4 X2X4 RES=U
OU
```

The estimated equation is

```
Y =  0.936   + 0.340*X1  + 0.399*X3  + 0.965*X1X3 + Error,  R² = 0.594
    (0.0504)  (0.115)     (0.0883)    (0.164)
    18.569    2.948       4.516       5.899
```

The raw data on all the variables, including the residual u, may be saved in a file by adding RA=KJRES.RAW on the OU command. Once this has been done, one can analyze these data and verify that indeed u is independent of x_2, x_4, and $x_2 x_4$ but not independent of x_1, x_3, and $x_1 x_3$. This is seen by running the following PRELIS command file (see KJTSLS2.PR2):

```
Checking the Residual
DA NI=10
LA; Y X1 X2 X3 X4 X1X3 X1X4 X2X3 X2X4 U
RA=KJRES.RAW;  CO ALL
SD Y X1X4 X2X3
RG U ON X1 X3 X1X3
RG U ON X2 X4 X2X4
OU MA=KM
```

3.7.2 Estimation by Means of Latent Variable Scores

In this section, we illustrate yet another simple way of estimating the non-linear equation (3.28) in the Kenny–Judd model, namely by means of latent variable scores as described in Section 3.6.

The PSF file corresponding to KJUDD.RAW is KJUDD.PSF. This can be obtained by running the following PRELIS command file (see file: KJUDD.PR2):

```
Computing PSF file from KJUDD.RAW
DA NI=5
LA;       Y X1 X2 X3 X4
RA=KJUDD.RAW;  CO ALL
OU MA=CM RA=KJUDD.PSF
```

This run will also produce a DSF file called KJUDD.DSF. To obtain the latent variable scores for ξ_1 and ξ_2 in 3.29, use the data system file KJUDD.DSF and the following SIMPLIS command file (see file KENJUDD.SPL):

```
Estimating the Measurement Model in the Kenny-Judd Model
   and Latent Variable Scores
System File from File KJUDD.DSF
Latent Variables Ksi1 Ksi2
Relationships
X1=1*Ksi1
X2=Ksi1
X3=1*Ksi2
X4=Ksi2
PSFfile KJUDD.PSF
End of Problem
```

Alternatively, one can use the following LISREL command file (see file KENJUDD.LS8):

3.8 SIMULATION

```
Estimating the Measurement Model in the Kenny-Judd Model
   and the Latent Variable Scores
SY=KJUDD.DSF
SE;  2 3 4 5 /
MO NX=4 NK=2 LX=FU,FI PH=SY,FR TD=DI,FR
LK;  Ksi1 Ksi2
FR  LX(2,1) LX(4,2);  VA  1.00 LX(1,1) LX(3,2)
PS=KJUDD.PSF
OU
```

Verify that the PSF file KJUDD.PSF has been appended with the scores on ξ_1 and ξ_2. One can now estimate (3.28) directly using the following PRELIS (see file KENJUDD.PR2):

```
Estimating Kenny-Judd Model from Latent Variable Scores
SY =KJudd.PSF;  CO ALL
NE   Ksi1Ksi2 = Ksi1*Ksi2
RG   Y  ON  Ksi1 Ksi2 Ksi1Ksi2
OU
```

The equation is estimated as

```
Y = 1.082   + 0.232*Ksi1 + 0.290*Ksi2 + 0.431*Ksi1Ksi2
   (0.0207) (0.0297)      (0.0218)     (0.0261)
    52.196   7.814         13.281       16.540
```

3.8 Simulation

The use of PRELIS and LISREL in simulation studies is described on pp. 185–206 in Jöreskog & Sörbom (1996a). One can generate data from many distributions and study the behavior of parameter estimates, standard errors, t-values, fit statistics, and other quantities obtained by any method available in LISREL. Here we describe how to study the properties of TSLS estimates.

Suppose, we want to simulate the Kenny–Judd model and study the properties of the TSLS estimates of the parameters α, γ_1, γ_2, and γ_3 in the Kenny–Judd model, see Section 3.7. This can be done with PRELIS alone.

To generate data for the Kenny–Judd model, we make the following assumptions

- ξ_1 and ξ_2 are bivariate normal with zero means
- $\zeta \sim N(0, \psi)$
- $\delta_i \sim N(0, \theta_i)$, $i = 1, \ldots, 4$
- δ_i is independent of δ_j for $i \neq j$
- δ_i is independent of ξ_j for $i = 1, \ldots, 4$ and $j = 1, 2$
- ζ is independent of δ_i and ξ_j for $i = 1, \ldots, 4$ and $j = 1, 2$

Although we are generating normal variables, the product variables used in the analysis are not normal. And although the TSLS approach is only estimating the four parameters α, γ_1, γ_2, and γ_3 in 3.28, there are 18 parameters in the Kenny–Judd model, namely, $\gamma_1, \gamma_2, \gamma_3, \lambda_2, \lambda_4, \phi_{11}, \phi_{21}, \phi_{22}, \psi$, $\theta_1, \theta_2, \theta_3, \theta_4, \alpha$ and $\tau_i, i = 1, 2, 3, 4$, which must all be specified to be able to generate data on the observed variables. We have chosen the following population parameter values:

$$\gamma_1 = 0.2, \quad \gamma_2 = 0.4, \quad \gamma_3 = 0.7, \quad \lambda_2 = 0.6, \quad \lambda_4 = 0.7,$$

$$\phi_{11} = 0.49, \quad \phi_{21} = 0.2352, \quad \phi_{22} = 0.64, \quad \psi = 0.20,$$

$$\theta_1 = 0.51, \quad \theta_2 = 0.64, \quad \theta_3 = 0.36, \quad \theta_4 = 0.51,$$

$$\alpha = 1.00, \quad \tau_i = 0.00, \quad i = 1, 2, 3, 4.$$

One can generate the data and estimate the model at the same time. This can be done with the following PRELIS command file (see file KJSIM.PR2; here it is assumed that the sample size is 800 and number of replicates is 600):

```
Generating Data for Kenny-Judd Model
and Estimating the Model at the Same Time
Simulating 600 Replicates
DA NO=800 RP=600
CO ALL
NE KSI1=.7*NRAND
NE KSI2=.48*KSI1+.726019*NRAND
NE Y=1+.2*KSI1+.4*KSI2+.7*KSI1*KSI2+.447214*NRAND
NE X1=KSI1+.7141428*NRAND
NE X2=.6*KSI1+.8*NRAND
NE X3=KSI2+.6*NRAND
NE X4=.7*KSI2+.7141428*NRAND
NE X1X3=X1*X3
```

3.8 SIMULATION

```
NE X1X4=X1*X4
NE X2X3=X2*X3
NE X2X4=X2*X4
NE ONE=1
SD KSI1 KSI2
RG Y on ONE X1 X2 X1X2 with ONE X3 X4 X3X4
OU IX=5967492 MA=MM YE=KJ.EST YS=KJ.STE XO
```

This will save 600 estimates of the four parameters in the file KJ.EST and the 600 corresponding standard errors in the file KJ.STE. These files can be read by PRELIS in free format to study the distribution of the estimates and the standard errors. Typically, one is interested in the bias in parameter estimates, *i.e.*, the mean of the estimates in KJ.EST minus the true parameter values used to generate the data and in the bias in standard errors, *i.e.*, the mean of the standard errors in KJ.STE minus the standard deviation of the parameter estimates in KJ.EST.

One problem with simulation studies like this is that they generate a very large output file which is not of any interest. To avoid this problem put XO on the OU command. This will produce output only for the first replicate. If one wants to see the output for the first four replicates, say, put XO=4 instead.

4   Standard Errors and Chi-Squares

LISREL[1] can compute two different sets of standard errors of parameter estimates and up to four different chi-squares for testing overall fit of the model. These new standard errors and chi-squares can be obtained for single group problems as well as multiple group problems and for covariance structure models in which only covariance matrices are analyzed, as well as for mean and covariance structure models, where both means and covariance matrices are analyzed.

Which set of standard errors and which chi-squares will be obtained depends on whether an asymptotic covariance matrix is provided or not and which method of estimation (ULS, GLS, ML, WLS, DWLS) is used to fit the model. The asymptotic covariance matrix is a consistent estimate of N times the asymptotic covariance matrix of the matrix being analyzed. This is computed by PRELIS and saved in a binary file which is read by LISREL, see Examples that follow.

4.1   Standard Errors

If no asymptotic covariance matrix is provided, standard errors are estimated under multivariate normality. Otherwise, if an asymptotic covariance matrix is provided, standard errors are estimated under non-normality.

---

[1] Introduced with version 8.20. Appendix A has a highly technical discussion about the formulas used in the computation of robust standard errors and chi-squares.

4.2 Chi-squares

The four different chi-squares are denoted C1, C2, C3, C4. Which ones are obtained in different situations is seen in the following table, where ● means obtained and ○ means not obtained (⋆ means that C2 will be obtained only if the asymptotic variances are provided).

Asymptotic Covariance Matrix not Provided

	ULS	GLS	ML	WLS	DWLS
C1	○	●	●	○	○
C2	●	●	●	○	⋆
C3	○	○	○	○	○
C4	○	○	○	○	○

Asymptotic Covariance Matrix Provided

	ULS	GLS	ML	WLS	DWLS
C1	○	●	●	●	○
C2	●	●	●	○	●
C3	●	●	●	○	●
C4	●	●	●	○	●

- C1 is $N-1$ times the minimum value of the fit function.
- C2 is $N-1$ times the minimum of the WLS fit function using a weight matrix estimated under multivariate normality.
- C3 is the Satorra-Bentler scaled chi-square statistic (Satorra & Bentler, 1988, equation 4.1) or its generalization to mean and covariance structures and multiple groups.
- C4 is computed from equation (2.20a) in Browne (1984) using the asymptotic covariance matrix provided, and in more general cases, from equation (30) in Satorra (1993).

Under multivariate normality of the observed variables, C1 and C2, whenever provided, are asymptotically equivalent and have an asymptotic chi-square distribution if the model holds exactly and an asymptotic noncentral chi-square distribution if the model holds approximately. The same holds for C4 under the more general assumption that the observed variables have a multivariate distribution with finite moments up to order four. C3 is a correction to C2 which makes C3 have the correct asymptotic mean even under non-normality. This correction is applied to C2, not to C1.

Even though C2 and C4 are correct asymptotic chi-squares under normality and non-normality, respectively, it does not mean that these are the "best" chi-squares in small and moderate samples. By Monte Carlo simulations, Hu, Bentler, & Kano (1992) found that, for a certain type and size of model, C3 performed better overall over a number of different sample sizes and degrees of non-normality. Further studies are needed to determine the relative advantages and disadvantages of these and other chi-square statistics in small and moderate samples, under different types and sizes of models, and under different distributions of the observed variables, see also Yuan & Bentler (1997).

4.3 LISREL implementation

If no asymptotic covariance matrix is provided and no method of estimation is specified, LISREL will use ML by default and estimate standard errors and C1 and C2 assuming that the observed variables are multivariate normal.

If an asymptotic covariance matrix is provided and no method of estimation is specified, LISREL will use WLS (ADF) by default and estimate standard errors and C1, under non-normality.

In both cases, any other method of estimation than the default may be specified, in which case, the standard errors and chi-squares are estimated under non-normality (C3 and C4) or normality (C1 and C2), depending on whether an asymptotic covariance matrix is provided or not. For ULS and DWLS, C1 does not have an asymptotic chi-square distribution, so for these methods, C1 is not given.

In an input file in the SIMPLIS command language, the method of estimation may be specified as (here exemplified by ULS)

```
Method: Unweighted Least Squares
```

or by

```
Options: ULS
```

or by

```
LISREL Output: ULS
```

In an input file in the LISREL command language, simply write ME=ULS on the OU command.

An asymptotic covariance matrix may be provided by writing

```
Asymptotic Covariance Matrix from File [filename]
```

in a SIMPLIS command file, or by writing

```
AC=[filename]
```

in a LISREL command file.

4.4 Other Fit Statistics

In LISREL 8, starting with version 8.20, there are many fit statistics other than the chi-square and its associated degrees of freedom and p-value. Many of these depend on chi-square explicitly or implicitly, such as NCP and RMSEA and their confidence limits. As there are now several chi-squares available, a decision has been made to base these other fit statistics on C2 in case of normality and C3 in case of non-normality. This makes a difference compared to previous versions of LISREL, where C1 was used as a basis for computing the other fit statistics. The reason for choosing C2 and C3 is that these are available for all methods except WLS.

4.5 GF File

The GF file is a file containing all the fit statistics including all the lower and upper confidence limits. It is described in Jöreskog & Sörbom (1996a, pp. 194–195). The numbers in the GF file are given in the same order as in the output file. The GF file is useful only in simulation studies (Bootstrap or Monte Carlo) when one wants to study the distribution of these fit statistics over a number of replicates. The GF file is obtained by writing

Options: GF=*[filename]*

in a SIMPLIS command file or by writing

GF=*[filename]*

on the OU command in a LISREL command file.

There is a short version of the GF file listing only the degrees of freedom, the chi-square, and the p-value for chi-square. This will be obtained if the keyword XI is also specified.

The difference between LISREL, version 8.20 and later, and previous versions is that there are now four chi-squares, so the GF file now contains four chi-squares and their corresponding p-values. These are given in the order C1, P1, C2, P2, C3, P3, C4, P4. Whenever a quantity has not been computed, a zero entry will appear in the GF file.

4.6 Examples

Two examples are given, one single group example, where the model fits the data well and one multi-group example where the model does not fit the data. This gives some "feeling" for how close the different chi-squares can be in these different situations.

Holzinger & Swineford (1939) collected data on twenty-six psychological tests administered to seventh- and eighth-grade children in two schools in Chicago: the Pasteur School and the Grant–White School. Six of these tests are selected for this example. The six tests are (with the original variable number in parenthesis):

VIS PERC	Visual Perception (V1)
CUBES	Cubes (V2)
LOZENGES	Lozenges (V4)
PAR COMP	Paragraph Comprehension (V6)
SEN COMP	Sentence Completion (V7)
WORDMEAN	Word meaning (V9)

The raw data on these six variables, in free format, is in the file SPV.RAW, where the first 156 cases constitute the Pasteur school sample and the last 145 cases constitute the Grant-White school sample.

The following stacked PRELIS input file computes the mean vector (ME), covariance matrix (CM), and asymptotic covariance matrix (AC) for each school and saves these in files with suffices ME, CM, and ACC, respectively (SPV.PR2).

```
Pasteur School
Data Ninputvars=6
Rawdata=SPV.RAW Rewind
Labels
'VIS PERC'  CUBES      LOZENGES
'PAR COMP' 'SEN COMP'  WORDMEAN
Continuous All
SCases Case < 157
Output Matrix=CMatrix ME=SPVPA.ME CM=SPVPA.CM AC=SPVPA.ACC

Grant-White School
Data Ninputvars=6
Rawdata=SPV.RAW
Labels
'VIS PERC'  CUBES      LOZENGES
'PAR COMP' 'SEN COMP'  WORDMEAN
Continuous All
SCases Case > 156
Output Matrix=CMatrix ME=SPVGW.ME CM=SPVGW.CM AC=SPVGW.ACC
```

The output file reveals some non-normality in both schools, in particular in variables LOZENGES and WORDMEAN:

```
The following lines were read from file L:\LISREL83\PR2EX\SPV.PR2:

Pasteur School
Data Ninputvars=6
Rawdata=SPV.RAW Rewind
Labels
'VIS PERC'  CUBES      LOZENGES
'PAR COMP' 'SEN COMP'  WORDMEAN
Continuous All
SCases Case < 157
Output Matrix=CMatrix ME=SPVPA.ME CM=SPVPA.CM AC=SPVPA.ACC

Total Sample Size =     156
```

[Output omitted]

4.6 EXAMPLES

Test of Univariate Normality for Continuous Variables

	Skewness		Kurtosis		Skewness and Kurtosis	
Variable	Z-Score	P-Value	Z-Score	P-Value	Chi-Square	P-Value
VISPERC	-1.955	0.051	1.682	0.093	6.652	0.036
CUBES	2.551	0.011	0.782	0.434	7.121	0.028
LOZENGES	1.155	0.248	-5.966	0.000	36.921	0.000
PARCOMP	1.455	0.146	0.018	0.985	2.117	0.347
SENCOMP	-1.020	0.308	-3.585	0.000	13.895	0.001
WORDMEAN	2.922	0.003	3.207	0.001	18.823	0.000

Relative Multivariate Kurtosis = 1.050

Test of Multivariate Normality for Continuous Variables

Skewness			Kurtosis			Skewness and Kurtosis	
Value	Z-Score	P-Value	Value	Z-Score	P-Value	Chi-Square	P-Value
5.141	5.403	0.000	3.028	1.779	0.075	32.357	0.000

```
Grant-White School
Data Ninputvars=6
Rawdata=SPV.RAW
Labels
'VIS PERC' CUBES LOZENGES 'PAR COMP' 'SEN COMP' WORDMEAN
Continuous All
SCases Case > 156
Output Matrix=CMatrix ME=SPVGW.ME CM=SPVGW.CM AC=SPVGW.ACC
```

Total Sample Size =   145

[Output omitted]

Test of Univariate Normality for Continuous Variables

	Skewness		Kurtosis		Skewness and Kurtosis	
Variable	Z-Score	P-Value	Z-Score	P-Value	Chi-Square	P-Value
VISPERC	-0.589	0.556	0.045	0.964	0.349	0.840
CUBES	1.183	0.237	1.843	0.065	4.795	0.091
LOZENGES	3.086	0.002	-1.320	0.187	11.263	0.004
PARCOMP	2.003	0.045	0.761	0.447	4.593	0.101
SENCOMP	-2.723	0.006	0.693	0.489	7.892	0.019
WORDMEAN	3.608	0.000	0.720	0.472	13.533	0.001

Relative Multivariate Kurtosis = 1.064

```
Test of Multivariate Normality for Continuous Variables

         Skewness                    Kurtosis          Skewness and Kurtosis

 Value   Z-Score  P-Value    Value   Z-Score  P-Value    Chi-Square  P-Value
 ------  -------  -------    ------  -------  -------    ----------  -------
 3.974    3.190    0.001     3.716    2.041    0.041       14.339     0.001
```

Since the sample sizes are rather small, it may not be a good idea to use the WLS (ADF) method; it may be better to take non-normality into account by means of C3 or C4

The following SIMPLIS input file (SPV1.SPL) estimates and tests a confirmatory factor analysis model with two correlated factors: Visual (Visual Perception) and Verbal (Verbal Ability), where the first three variables are indicators of Visual and the last three variables are indicators of Verbal. The model is estimated by maximum likelihood but standard errors are estimated under non-normality.

```
Estimating and Testing a Confirmatory Factor Analysis Model
on Grant-White School

Observed Variables: 'VIS PERC' CUBES LOZENGES 'PAR COMP' 'SEN COMP' WORDMEAN
Covariance Matrix from File SPVGW.CM
Asymptotic Covariance Matrix from File SPVGW.ACC
Sample Size: 145
Latent Variables: Visual Verbal
Relationships:
'VIS PERC' - LOZENGES Visual
'PAR COMP' - WORDMEAN Verbal
Method: Maximum Likelihood
End of Problem
```

First run this model assuming normality, *i.e.*, without the line

```
Asymptotic Covariance Matrix from File SPVGW.ACC
```

This can be done by putting an exclamation sign ! in front of the word 'Asymptotic.'

The following solution is obtained:

4.6 EXAMPLES

```
LISREL Estimates (Maximum Likelihood)

VIS PERC = 4.37*Visual, Errorvar.= 28.71, R² = 0.40
          (0.64)                  (4.79)
           6.78                    6.00

   CUBES = 2.37*Visual, Errorvar.= 14.15, R² = 0.28
          (0.41)                  (2.00)
           5.71                    7.08

LOZENGES = 6.09*Visual, Errorvar.= 32.12, R² = 0.54
          (0.79)                  (7.35)
           7.74                    4.37

PAR COMP = 2.93*Verbal, Errorvar.= 2.81 , R² = 0.75
          (0.24)                  (0.59)
          12.35                    4.76

SEN COMP = 3.83*Verbal, Errorvar.= 6.92 , R² = 0.68
          (0.33)                  (1.18)
          11.49                    5.88

WORDMEAN = 6.58*Verbal, Errorvar.= 19.83, R² = 0.69
          (0.57)                  (3.42)
          11.56                    5.80


        Correlation Matrix of Independent Variables

                 Visual      Verbal
                 --------    --------
       Visual     1.00

       Verbal     0.53        1.00
                 (0.09)
                  6.23
```

This solution gives the chi-squares C1 and C2 as follows:

```
                    Goodness of Fit Statistics

                      Degrees of Freedom = 8
             Minimum Fit Function Chi-Square = 3.64 (P = 0.89)
       Normal Theory Weighted Least Squares Chi-Square = 3.70 (P = 0.88)
```

Next run the problem under non-normality (SPV2.SPL), *i.e.*, including the line:

Asymptotic Covariance Matrix from File SPVGW.ACC

This gives the following solution, where the standard error have been estimated under non-normality:

```
LISREL Estimates (Maximum Likelihood)

 VIS PERC = 4.37*Visual, Errorvar.= 28.71, R² = 0.40
           (0.72)                  (6.33)
            6.06                    4.54

    CUBES = 2.37*Visual, Errorvar.= 14.15, R² = 0.28
           (0.37)                  (2.27)
            6.40                    6.22

 LOZENGES = 6.09*Visual, Errorvar.= 32.12, R² = 0.54
           (0.79)                  (8.13)
            7.66                    3.95

 PAR COMP = 2.93*Verbal, Errorvar.= 2.81 , R² = 0.75
           (0.25)                  (0.60)
           11.62                    4.72

 SEN COMP = 3.83*Verbal, Errorvar.= 6.92 , R² = 0.68
           (0.33)                  (1.19)
           11.51                    5.83

 WORDMEAN = 6.58*Verbal, Errorvar.= 19.83, R² = 0.69
           (0.58)                  (3.74)
           11.42                    5.30


         Correlation Matrix of Independent Variables

                  Visual       Verbal
                --------     --------
       Visual      1.00

       Verbal      0.53         1.00
                  (0.09)
                   5.74
```

In this case, all four chi-squares C1, C2, C3, C4 are given in the output in that order:

```
                      Goodness of Fit Statistics

                       Degrees of Freedom = 8
            Minimum Fit Function Chi-Square = 3.64 (P = 0.89)
    Normal Theory Weighted Least Squares Chi-Square = 3.70 (P = 0.88)
            Satorra-Bentler Scaled Chi-Square = 3.94 (P = 0.86)
            Chi-Square Corrected for Non-Normality = 4.12 (P = 0.85)
```

4.6 EXAMPLES

This may be compared with what was obtained under normality. Note that the parameter estimates are the same (they have been estimated by ML in both cases) but the standard errors and t-values are slightly different. This is the effect of non-normality. Also note, that the two chi-squares C1 and C2 are the same as in the previous case.

The following table shows the various chi-square values that are obtained with different methods, assuming the asymptotic covariance matrix is provided. The degrees of freedom is 8.

	ULS	GLS	ML	WLS	DWLS
C1	o	3.43	3.64	4.13	o
C2	3.70	3.87	3.70	o	3.66
C3	4.03	3.97	3.94	o	3.85
C4	4.13	4.12	4.12	o	4.07

Continuing the example, the following input file (SPV3.SPL) estimates the same confirmatory factor analysis model simultaneously in both schools assuming that the intercepts and the factor loadings are the same in the two groups and estimates the mean difference in the latent variables between schools, see Sörbom (1974).

```
Group: Grant-White
Estimating Latent Mean Difference between Two Schools
Observed Variables: 'VIS PERC' CUBES LOZENGES 'PAR COMP' 'SEN COMP' WORDMEAN
Means from File SPVGW.ME
Covariance Matrix from File SPVGW.CM
Asymptotic Covariance Matrix from File SPVGW.ACC
Sample Size: 145
Latent Variables: Visual Verbal
Relationships:
'VIS PERC' - LOZENGES = CONST Visual
'PAR COMP' - WORDMEAN = CONST Verbal
'VIS PERC' = 1*Visual
'PAR COMP' = 1*Verbal
Method: Maximum Likelihood

Group: Pasteur
Estimating Latent Mean Difference between Two Schools
Means from File SPVPA.ME
Covariance Matrix from File SPVPA.CM
Asymptotic Covariance Matrix from File SPVPA.ACC
```

```
Sample Size: 156
Relationships:
Visual - Verbal = CONST
Set the Variances of Visual - Verbal Free
Set the Covariance of Visual - Verbal Free
Set the Error Variances of 'VIS PERC' - WORDMEAN Free
End of Problem
```

The corresponding chi-squares are now given in the output as:

```
                        Goodness of Fit Statistics

                          Degrees of Freedom = 24
                 Minimum Fit Function Chi-Square = 63.39 (P = 0.00)
        Normal Theory Weighted Least Squares Chi-Square = 60.27 (P = 0.00)
               Satorra-Bentler Scaled Chi-Square = 66.35 (P = 0.00)
              Chi-Square Corrected for Non-Normality = 71.45 (P = 0.00)
```

It is clear that this model does not fit the data. Further study of the output reveals that the intercept for LOZENGES is probably different in the two schools. Even though the model does not fit, the output suggests that Grant–White School students are ahead of the Pasteur School students in verbal ability. This agrees with the results found and interpretations made by Jöreskog & Sörbom (1996c, pp. 75–76).

The following table shows the various chi-square values that are obtained with different methods, assuming the asymptotic covariance matrices are provided. The degrees of freedom is 24.

	ULS	GLS	ML	WLS	DWLS
C1	o	61.35	63.39	960.63	o
C2	54.81	74.74	60.27	o	58.87
C3	61.88	72.25	66.35	o	61.48
C4	71.81	76.03	71.45	o	67.30

The solutions for WLS and DWLS are non-admissible. This shows that using asymptotic variances and covariances estimated from a small sample can do more harm than good if used with WLS or DWLS. In such cases it is probably better to use any of the other methods and correct for non-normality with C3 or C4.

A Robust Standard Errors and Chi-Squares

This technical appendix describes how we implemented the computation of robust standard errors and chi-squares in LISREL.[1]

Several authors have contributed to the statistical inference theory for covariance structures in single and multiple groups, notably Browne (1977, 1984, 1987), Jöreskog (1981), and Satorra (1987, 1993). The purpose of this appendix is *not* to review this literature, but rather to give the formulas we use in LISREL to compute standard errors and chi-squares.

A non-technical description of the standard errors and chi-squares is given in Chapter 4 *Standard Errors and Chi-Squares*, starting on page 179.

It should be noted that these formulas are for continuous variables. The theory for ordinal variables is another matter.

1. Single Group: Covariance Structures
2. Single Group: Mean and Covariance Structures
3. Single Group: Augmented Moment Matrices
4. Multiple Groups: Covariance Structures
5. Multiple Groups: Mean and Covariance Structures
6. Multiple Groups: Augmented Moment Matrices

---

[1] Written by Karl G. Jöreskog & Dag Sörbom

APPENDIX A: ROBUST STANDARD ERRORS AND CHI-SQUARES

A.1 Single Group: Covariance Structures

Definitions

$$k = \text{number of observed variables} \quad (A.1)$$

$$s = \frac{1}{2}k(k+1) \quad (A.2)$$

$$t = \text{number of independent parameters} < s \quad (A.3)$$

$$d = s - t \quad (A.4)$$

$$\underset{k^2 \times s}{\mathbf{K}} = \mathbf{D}(\mathbf{D}'\mathbf{D})^{-1}, \text{ where } \underset{k^2 \times s}{\mathbf{D}} \text{ is the duplication matrix} \quad (A.5)$$

$$\underset{s \times 1}{\mathbf{s}} = \mathbf{K}'\text{vec}(\mathbf{S}) \qquad \text{vec}(\mathbf{S}) = \mathbf{D}\mathbf{s} \quad (A.6)$$

$$\underset{s \times s}{\boldsymbol{\Omega}} = n\text{ACov}(\mathbf{s}) \qquad n = N - 1 \quad (A.7)$$

$$\underset{s \times s}{\mathbf{W}} = n\,\text{Est}[\text{ACov}(\mathbf{s})] \quad (A.8)$$

$$\mathbf{W}_{\text{NT}} = 2\mathbf{K}'(\hat{\boldsymbol{\Sigma}} \otimes \hat{\boldsymbol{\Sigma}})\mathbf{K} \qquad \text{under NT, } i.e., \text{ if AC not read} \quad (A.9)$$

$$\mathbf{W}_{\text{NNT}} = \text{computed by PRELIS, } i.e., \text{ if AC read by LISREL} \quad (A.10)$$

$$\boldsymbol{\sigma} = \underset{s \times 1}{\boldsymbol{\sigma}}(\underset{t \times 1}{\boldsymbol{\theta}}) \quad (A.11)$$

$$\underset{s \times t}{\boldsymbol{\Delta}} = \frac{\partial \boldsymbol{\sigma}}{\partial \boldsymbol{\theta}'} \text{ evaluated at } \hat{\boldsymbol{\theta}} \text{ and assumed to have rank } t \quad (A.12)$$

$$\underset{s \times d}{\boldsymbol{\Delta}_c} = \text{orthogonal complement to } \boldsymbol{\Delta} \quad (A.13)$$

$$\boldsymbol{\Delta}_c'\boldsymbol{\Delta} = 0 \qquad [\boldsymbol{\Delta}|\boldsymbol{\Delta}_c] \text{ non-singular}$$

Comments:

- In LISREL notation $k = p + q$ where p and q are the number of y- and x-variables, respectively.
- s is the number of independent elements of the sample covariance matrix $\mathbf{S}$.
- The assumption $t < s$ implies that we exclude the case of saturated models where $t = s$. For such models all residuals and chi-squares are zero. The asymptotic covariance matrix of parameter estimates

A.1 SINGLE GROUP: COVARIANCE STRUCTURES

can still be obtained by (A.24) but some of the other formulas break down. For example, Δ_c does not exist.

- We assume that the model is identified so that d is the degrees of freedom of the model.

- vec(S) is a vector of order $k^2 \times 1$ consisting of the columns of S stringed under each other.

- $\mathbf{s} = (s_{11}, s_{21}, s_{22}, s_{31}, ..., s_{kk})'$ is a vector of order $s \times 1$ consisting of the nonduplicated elements of S. D is the duplication matrix (see Magnus & Neudecker, 1988) which transforms s to vec(S). $\mathbf{K}'$ is the generalized inverse of D, which transforms vec(S) to s.

- Ω is n times the asymptotic covariance matrix of s. This is unknown but can be estimated by W. The estimated asymptotic covariance matrix W is abbreviated AC.

- Under normal theory NT, i.e., if the observations come from a multivariate normal population or if S has a Wishart distribution, W can be estimated by $\mathbf{W}_{NT}$ in (A.9). Here $\otimes$ is the symbol for a Kronecker product. LISREL uses this formula to estimate W if no AC matrix is read. In (A.9), $\hat{\Sigma}$ is the fitted covariance matrix obtained after the model has been estimated.

- In the non-normal case NNT, i.e., if the observations come from a distribution which is not multivariate normal, the elements of W can be estimated by the formula

$$w_{gh,ij} = n\text{Est}[\text{ACov}(s_{gh}, s_{ij})] = m_{ghij} - s_{gh}s_{ij}, \quad (\text{A.14})$$

where

$$m_{ghij} = (1/N) \sum_{a=1}^{N} (z_{ag} - \bar{z}_g)(z_{ah} - \bar{z}_h)(z_{ai} - \bar{z}_i)(z_{aj} - \bar{z}_j) \quad (\text{A.15})$$

is a fourth-order central sample moment. This defines $\mathbf{W}_{NNT}$. PRELIS computes $\mathbf{W}_{NNT}$ for continuous variables and LISREL will use this matrix if it is read.

- The covariance structure model is $\Sigma = \Sigma(\theta)$, i.e., the elements of Σ are functions of the parameters θ. In terms of the nonduplicated elements of Σ, this is expressed as in (A.11).

- $\boldsymbol{\Delta}$ is a matrix of derivatives. It can be evaluated at any point in the parameter space. However, its operational use is when it is evaluated at the parameter estimates, and this is assumed in what follows.
- Since $\boldsymbol{\Delta}$ is of rank $t < s$ there exists $s - t = d$ columns orthogonal to the columns in $\boldsymbol{\Delta}$. These d columns form the orthogonal complement $\boldsymbol{\Delta}_c$. This is not unique but any orthogonal complement will do.

Fit functions

With these definitions, all fit functions in LISREL are of the same form:

$$F = (\mathbf{s} - \boldsymbol{\sigma})' \underset{s \times s}{\mathbf{V}} (\mathbf{s} - \boldsymbol{\sigma}),$$

where the weight matrix $\mathbf{V}$ is defined differently for different fit functions:

$$\text{ULS:} \quad \mathbf{V} = \mathbf{I}^* = \text{diag}(1, 2, 1, 2, 2, 1, \ldots) \tag{A.16}$$

$$\text{GLS:} \quad \mathbf{V} = \mathbf{D}'(\mathbf{S}^{-1} \otimes \mathbf{S}^{-1})\mathbf{D} \tag{A.17}$$

$$\text{ML:} \quad \mathbf{V} = \mathbf{D}'(\hat{\boldsymbol{\Sigma}}^{-1} \otimes \hat{\boldsymbol{\Sigma}}^{-1})\mathbf{D} \tag{A.18}$$

$$\text{WLS:} \quad \mathbf{V} = \mathbf{W}_{\text{NNT}}^{-1} \quad \text{or} \quad \mathbf{W}_{\text{NNT}}^{-} \quad \text{if } \mathbf{W}_{\text{NNT}} \text{ singular} \tag{A.19}$$

$$\text{DWLS:} \quad \mathbf{V} = \mathbf{D}_{\mathbf{W}}^{-1} = [\text{diag}\,\mathbf{W}]^{-1} \tag{A.20}$$

with
$$\mathbf{W} = \mathbf{W}_{\text{NT}} \quad \text{if AC not read}$$
$$\mathbf{W} = \mathbf{W}_{\text{NNT}} \quad \text{if AC read}$$

Comments:

- The fit function for ML is usually written

$$F = \log ||\boldsymbol{\Sigma}|| + \text{tr}(\mathbf{S}\boldsymbol{\Sigma}^{-1}) - \log ||\mathbf{S}|| - k \tag{A.21}$$

but an asymptotically equivalent form is

$$F = (\mathbf{s} - \boldsymbol{\sigma})'\mathbf{D}'(\hat{\boldsymbol{\Sigma}}^{-1} \otimes \hat{\boldsymbol{\Sigma}}^{-1})\mathbf{D}(\mathbf{s} - \boldsymbol{\sigma}) \tag{A.22}$$

A.1 SINGLE GROUP: COVARIANCE STRUCTURES

Equation (A.22) can be interpreted as ML estimated by means of iteratively reweighted least squares in which $\hat{\Sigma}$ is updated in each iteration. Both of these fit functions have a minimum at the same point in the parameter space, namely at the ML estimates. However, the minimum value of the functions are not the same.

Results

$$\underset{s \times s}{\mathbf{E}} = \mathbf{\Delta}'\mathbf{V}\mathbf{\Delta} \tag{A.23}$$

$$n\mathrm{ACov}(\hat{\theta}) = \mathbf{E}^{-1}\mathbf{\Delta}'\mathbf{VWV}\mathbf{\Delta}\mathbf{E}^{-1} \tag{A.24}$$

$$n\mathrm{ACov}(\mathbf{s} - \hat{\sigma}) = \mathbf{W} - \mathbf{\Delta}\mathbf{E}^{-1}\mathbf{\Delta}' \tag{A.25}$$

$$\text{with } \mathbf{W} = \mathbf{W}_{\mathrm{NT}} \quad \text{if AC not read} \tag{A.26}$$

$$\mathbf{W} = \mathbf{W}_{\mathrm{NNT}} \quad \text{if AC read} \tag{A.27}$$

$$c_1 = n \min F \quad \text{for ULS, GLS, WLS, DWLS} \tag{A.28}$$

$$= n\left(\log ||\hat{\Sigma}|| + \mathrm{tr}(\mathbf{S}\hat{\Sigma}^{-1}) - \log ||\mathbf{S}|| - k\right) \quad \text{for ML}$$

$$c_2 = n(\mathbf{s} - \hat{\sigma})'\mathbf{\Delta}_c(\mathbf{\Delta}'_c\mathbf{W}_{\mathrm{NT}}\mathbf{\Delta}_c)^{-1}\mathbf{\Delta}'_c(\mathbf{s} - \hat{\sigma}) \tag{A.29}$$

$$h_1 = \mathrm{tr}[(\mathbf{\Delta}'_c\mathbf{W}_{\mathrm{NT}}\mathbf{\Delta}_c)(\mathbf{\Delta}'_c\mathbf{W}_{\mathrm{NNT}}\mathbf{\Delta}_c)] \tag{A.30}$$

$$c_3 = \frac{d}{h_1}c_2 \tag{A.31}$$

$$c_4 = n(\mathbf{s} - \hat{\sigma})'\mathbf{\Delta}_c(\mathbf{\Delta}'_c\mathbf{W}_{\mathrm{NNT}}\mathbf{\Delta}_c)^{-1}\mathbf{\Delta}'_c(\mathbf{s} - \hat{\sigma}) \tag{A.32}$$

Comments:

- $\mathbf{E}$ is the information matrix.
- (A.24) gives the estimated asymptotic covariance matrix of the parameter estimates. The standard errors of the parameter estimates are obtained from the diagonal elements of this matrix.
- Note that if AC is not read and ML is used then $\mathbf{W} = \mathbf{V}^{-1}$ and (A.24) reduces to

$$n\mathrm{ACov}(\hat{\theta}) = \mathbf{E}^{-1} \tag{A.33}$$

- Similarly if AC is read and WLS is used, then $\mathbf{VWV} = \mathbf{V}$ and then also (A.24) reduces to (A.33).

- The standard errors of the residuals are obtained from the diagonal elements of (A.25). From these the standardized residuals are obtained.

- c_1, c_2, c_3, and c_4, define four different chi-squares

    - c_1 is n times the minimum value of the fit function.
    - c_2 is n times the minimum of the WLS fit function using a weight matrix estimated under multivariate normality.
    - c_3 is the Satorra–Bentler scaled chi-square statistic (Satorra & Bentler, 1988, equation 4.1)
    - c_4 is equation (2.20a) in Browne (1984) using the asymptotic covariance matrix provided.
    - Under multivariate normality of the observed variables, c_1 and c_2 are asymptotically equivalent and have an asymptotic chi-square distribution if the model holds exactly and an asymptotic non-central chi-square distribution if the model holds approximately. The same holds for c_4 under the more general assumption that the observed variables have a multivariate distribution with finite moments up to order four. c_3 is a correction to c_2 which makes c_3 have the correct asymptotic mean even under non-normality. This correction is applied to c_2, not to c_1.

- Note that the inverse of $\mathbf{W}$ is not required in any of these formulas, except if WLS is used. This is important because the inverse of $\mathbf{W}$ is often more unstable than $\mathbf{W}$ itself.

A.2 Single Group: Mean and Covariance Structures

There are two possible approaches to analyze the covariance matrix and the mean vector jointly. One is described in this section and the other in the next section.

In principle, the approach of the previous section can be applied to any vector of moments. So we just extend the vector $\mathbf{s}$ with the mean vector $\bar{\mathbf{z}}$ and then use the same definitions as before:

A.2 SINGLE GROUP: MEAN AND COVARIANCE STRUCTURES

$$s = \frac{1}{2}k(k+1) + k \quad (A.34)$$

$$t = \text{as before} \quad (A.35)$$

$$\beta(\theta) = \begin{bmatrix} \sigma(\theta) \\ \mu(\theta) \end{bmatrix} \quad (A.36)$$

$$\Delta = \frac{\partial \beta}{\partial \theta'} \quad (A.37)$$

$$\Delta_c = \text{as before} \quad (A.38)$$

$$\mathbf{b} = \begin{bmatrix} \mathbf{s} \\ \bar{\mathbf{z}} \end{bmatrix} \quad (A.39)$$

Ω is now n times the asymptotic covariance matrix of $\mathbf{b}$ and $\mathbf{W}$ a consistent estimate of it. We assume that $\mathbf{s}$ and $\bar{\mathbf{z}}$ are asymptotically independent so that Ω and $\mathbf{W}$ are block diagonal.

Under normal theory NT, $\mathbf{W}$ is estimated as

$$\mathbf{W}_{\text{NTE}} = \begin{bmatrix} \mathbf{W}_{\text{NT}} & \\ 0 & \hat{\Sigma} \end{bmatrix} \quad (A.40)$$

and under non-normal theory NNT, $\mathbf{W}$ is estimated as

$$\mathbf{W}_{\text{NNTE}} = \begin{bmatrix} \mathbf{W}_{\text{NNT}} & \\ 0 & \hat{\Sigma} \end{bmatrix} \quad (A.41)$$

where $\mathbf{W}_{\text{NT}}$ and $\mathbf{W}_{\text{NNT}}$ are defined as in the previous section.

Fit functions

$$F = (\mathbf{b} - \beta)'\mathbf{V}(\mathbf{b} - \beta)$$

$$\text{ULS:} \quad \mathbf{V} = \begin{bmatrix} \mathbf{I}^* & \\ 0 & \hat{\Sigma}^{-1} \end{bmatrix} \quad (A.42)$$

$$\text{GLS:} \quad \mathbf{V} = \begin{bmatrix} \mathbf{D}'(\mathbf{S}^{-1} \otimes \mathbf{S}^{-1})\mathbf{D} & \\ 0 & \mathbf{S}^{-1} \end{bmatrix} \quad (A.43)$$

$$\text{ML:} \quad \mathbf{V} = \begin{bmatrix} \mathbf{D}'(\hat{\mathbf{\Sigma}}^{-1} \otimes \hat{\mathbf{\Sigma}}^{-1})\mathbf{D} & \\ 0 & \hat{\mathbf{\Sigma}}^{-1} \end{bmatrix} \quad (A.44)$$

$$\text{WLS:} \quad \mathbf{V} = \begin{bmatrix} \mathbf{W}_{\text{NNT}}^{-} & \\ 0 & \hat{\mathbf{\Sigma}}^{-1} \end{bmatrix} \quad (A.45)$$

$$\text{DWLS:} \quad \mathbf{V} = \begin{bmatrix} \mathbf{D}_{\mathbf{W}}^{-1} & \\ 0 & \hat{\mathbf{\Sigma}}^{-1} \end{bmatrix} \quad (A.46)$$

The same results as in Section A.1 apply but with $\mathbf{W}_{\text{NT}}$ replaced with $\mathbf{W}_{\text{NTE}}$ and $\mathbf{W}_{\text{NNT}}$ replaced with $\mathbf{W}_{\text{NNTE}}$.

A.3 Single Group: Augmented Moment Matrix

In the previous section it was assumed that $\mathbf{s}$ and $\bar{\mathbf{z}}$ are asymptotically independent. Under non-normality this may not hold. A way to avoid this assumption is to use the augmented moment matrix. This is the matrix of sample moments about zero for the vector $\mathbf{z}$ augmented with a variable which is constant equal to 1 for every case. The population augmented moment matrix and the sample augmented moment matrix are defined in equations (A.49) and (A.50) that follow. $\mathbf{a}$ is a vector of the non-duplicated elements $\mathbf{A}$. Because the last element of $\mathbf{a}$ is constant equal to 1, its covariance matrix $\mathbf{W}_a$ is singular. However the inverse of $\mathbf{W}_a$ is only used with WLS in which case it is replaced by its generalized inverse. Under non-normal theory $\mathbf{W}_a$ is estimated as $\mathbf{W}_{a\text{NNT}}$ whose elements are

$$w_{gh,ij} = n\text{Est}\left[\text{ACov}(a_{gh}, a_{ij})\right] = n_{ghij} - a_{gh}a_{ij}, \quad (A.47)$$

where

$$n_{ghij} = (1/N) \sum_{a=1}^{N} z_{ag} z_{ah} z_{ai} z_{aj} \quad (A.48)$$

is a fourth-order sample moment about zero.

Definitions

$$\Upsilon = \mathsf{E}\left[\begin{pmatrix} \mathbf{z} \\ 1 \end{pmatrix}\right][\mathbf{z}' \quad 1] = \begin{bmatrix} \Sigma + \mu\mu' & \mu \\ \mu' & 1 \end{bmatrix} \quad (A.49)$$

$$\mathbf{A} = \frac{1}{N}\sum_{c=1}^{N}\begin{bmatrix} \mathbf{z}_c \\ 1 \end{bmatrix}[\mathbf{z}'_c \quad 1] = \begin{bmatrix} \mathbf{S} + \bar{\mathbf{z}}\bar{\mathbf{z}}' & \bar{\mathbf{z}} \\ \bar{\mathbf{z}}' & 1 \end{bmatrix} \quad (A.50)$$

$$\mathbf{a} = \mathbf{K}'\text{vec}(\mathbf{A}) \quad \text{vec}(\mathbf{A}) = \mathbf{D}\mathbf{a} \quad (A.51)$$

$$\boldsymbol{\alpha} = \mathbf{K}'\text{vec}(\Upsilon) \quad (A.52)$$

$$\mathbf{W}_a = n\text{Est}[n\text{ACov}(\mathbf{a})] \quad \text{singular} \quad (A.53)$$

$$\mathbf{W}_{a\text{NT}} = 2\mathbf{K}'(\hat{\Upsilon} \otimes \hat{\Upsilon})\mathbf{K} \quad (A.54)$$

$$\mathbf{W}_{a\text{NNT}} = \text{computed by PRELIS, read by LISREL} \quad (A.55)$$

Fit functions

$$F = (\mathbf{a} - \boldsymbol{\alpha})'\mathbf{V}(\mathbf{a} - \boldsymbol{\alpha}) \quad (A.56)$$

with $\mathbf{V}$ defined as in Section A.1 but with

$\mathbf{A}$ instead of $\mathbf{S}$

Υ instead of Σ

The same results as in Section A.1 apply but with $\mathbf{W}_{\text{NT}}$ replaced with $\mathbf{W}_{a\text{NT}}$ and $\mathbf{W}_{\text{NNT}}$ replaced with $\mathbf{W}_{a\text{NNT}}$.

A.4 Multiple Groups: Covariance Structures

To generalize the results in Section A.1 to multiple groups we need only consider that the samples from different groups are supposed to be independent. The presentation here follows Satorra (1993).

The fit function in (A.59) is a weighted sum of the fit functions for each group. The asymptotic covariance matrix in (A.60) can be estimated for each group separately under NT and NNT as in Section A.1. The total asymptotic covariance matrix is defined in (A.61) and the **V**-matrix is defined in (A.63) where each $\mathbf{V}_g$ is n_g/n times the **V** in Section A.1. Δ is defined in (A.64). Δ_c is an orthogonal complement to Δ partitioned as in (A.65).

Definitions

$$\mathbf{s}' = (\mathbf{s}'_1, \mathbf{s}'_2, \ldots, \mathbf{s}'_G) \tag{A.57}$$

$$\boldsymbol{\sigma}' = (\boldsymbol{\sigma}'_1, \boldsymbol{\sigma}'_2, \ldots, \boldsymbol{\sigma}'_G) \tag{A.58}$$

$$n_g = N_g - 1, \qquad n = \sum n_g$$

$$F(\mathbf{s}, \boldsymbol{\sigma}) = \sum \frac{n_g}{n} F_g(\mathbf{s}_g, \boldsymbol{\sigma}_g) \tag{A.59}$$

$$\mathbf{W}_g = \mathrm{Est}[n_g \mathrm{ACov}(\mathbf{s}_g)] \tag{A.60}$$

$$\mathbf{W} = \begin{bmatrix} \mathbf{W}_1 & & & 0 \\ & \mathbf{W}_2 & & \\ & & \ddots & \\ 0 & & & \mathbf{W}_G \end{bmatrix} \quad \text{block diagonal} \tag{A.61}$$

$$\mathbf{V}_g = \frac{n_g}{n} \times \mathbf{V} \text{ defined in (A.16)–(A.20)} \tag{A.62}$$

different for different methods

$$\mathbf{V} = \begin{bmatrix} \mathbf{V}_1 & & & 0 \\ & \mathbf{V}_2 & & \\ & & \ddots & \\ 0 & & & \mathbf{V}_G \end{bmatrix} \quad \text{block diagonal} \tag{A.63}$$

A.4 MULTIPLE GROUPS: COVARIANCE STRUCTURES

$$\Delta = \frac{\partial \sigma}{\partial \theta'} = \begin{bmatrix} \frac{\partial \sigma_1}{\partial \theta'} \\ \frac{\partial \sigma_2}{\partial \theta'} \\ \vdots \\ \frac{\partial \sigma_G}{\partial \theta'} \end{bmatrix} = \begin{bmatrix} \Delta_1 \\ \Delta_2 \\ \vdots \\ \Delta_G \end{bmatrix} \quad (A.64)$$

$$\Delta_c = \begin{bmatrix} \Delta_{1c} \\ \Delta_{2c} \\ \vdots \\ \Delta_{Gc} \end{bmatrix} \quad (A.65)$$

Results

With these definitions, the same results as in Section A.1 apply, namely

$$\mathbf{E} = \Delta' \mathbf{V} \Delta \quad (A.66)$$
$$n\mathsf{ACov}(\theta) = \mathbf{E}^{-1} \Delta' \mathbf{V} \mathbf{W} \mathbf{V} \Delta \mathbf{E}^{-1} \quad (A.67)$$
$$n\mathsf{ACov}(\mathbf{s} - \hat{\sigma}) = \mathbf{W} - \Delta \mathbf{E}^{-1} \Delta' \quad (A.68)$$
$$c_1 = n \min F \quad (A.69)$$
$$c_2 = n(\mathbf{s} - \hat{\sigma})' \Delta_c (\Delta'_c \mathbf{W}_{\mathsf{NT}} \Delta_c)^{-1} \Delta'_c (\mathbf{s} - \hat{\sigma}) \quad (A.70)$$
$$h_1 = \mathrm{tr}[(\Delta'_c \mathbf{W}_{\mathsf{NT}} \Delta_c)(\Delta'_c \mathbf{W}_{\mathsf{NNT}} \Delta_c)] \quad (A.71)$$
$$c_3 = \frac{d}{h_1} c_2 \quad (A.72)$$
$$c_4 = n(\mathbf{s} - \hat{\sigma})' \Delta_c (\Delta'_c \mathbf{W}_{\mathsf{NNT}} \Delta_c)^{-1} \Delta'_c (\mathbf{s} - \hat{\sigma}) \quad (A.73)$$

but note that

$$\mathbf{E} = \Delta' \mathbf{V} \Delta = \sum_g \Delta'_g \mathbf{V}_g \Delta_g \quad (A.74)$$

$$\Delta' V W V \Delta = \sum_g \Delta'_g V_g W_g V_g \Delta_g \qquad (A.75)$$

$$n\text{ACov}(s_g - \hat{\sigma}_g) = W_g - \Delta_g E^{-1} \Delta'_g \qquad (A.76)$$

$$\Delta'_c W \Delta_c = \sum_g \Delta'_{gc} W_g \Delta_{gc} \qquad (A.77)$$

$$\Delta'_c (s - \hat{\sigma}) = \sum_g \Delta'_{gc} (s_g - \hat{\sigma}_g) \qquad (A.78)$$

A.5 Multiple Groups: Mean and Covariance Structures

The generalization of the results in Section A.2 to multiple groups follows by using a **b** vector instead of **s** in (A.59) and definitions in analogy to Sections A.2 and A.4.

A.6 Multiple Groups: Augmented Moment Matrices

The generalization of the results in Section A.3 to multiple groups follows by using an **a** vector instead of **s** in (A.59) and definitions in analogy to Sections A.3 and A.4.

B Why are t-Values for Error Variances Equal?

This technical appendix discusses a paradox in path models for observed variables: "Why are t-Values for Error Variances Equal?"[1]

Let $\mathbf{y} = (y_1, y_2, \ldots, y_p)$ be a set of jointly dependent (endogenous) variables and let $\mathbf{x} = (x_1, x_2, \ldots, x_q)$ be a set of independent (exogenous) variables. Consider any model of the form

$$\mathbf{y} = \boldsymbol{\alpha} + \mathbf{B}\mathbf{y} + \boldsymbol{\Gamma}\mathbf{x} + \mathbf{z}, \qquad (B.1)$$

where $\boldsymbol{\alpha} = (\alpha_1, \alpha_2, \ldots, \alpha_p)$ is a vector of intercept terms and $\mathbf{z} = (z_1, z_2, \ldots, z_p)$ is a vector of error terms assumed to be uncorrelated with $\mathbf{x}$. There are no latent variables in the model. The matrix $\mathbf{B}$ is assumed to have zero elements in and above the diagonal, i.e., (B.1) is a recursive system, see Jöreskog & Sörbom (1996b, pp. 143–145). Any element below the diagonal of $\mathbf{B}$ may be either a fixed zero or a free parameter. The matrix $\boldsymbol{\Gamma}$ may contain any number of fixed zeros and any number of free elements. The error covariance matrix $\boldsymbol{\Psi} = \mathrm{Cov}(\mathbf{z})$ is assumed to be diagonal. Suppose the model is estimated by ML from a sample of size N under the assumption that $\mathbf{z}$ is multivariate normal. Nothing needs to be assumed about the distribution of $\mathbf{x}$.

---

[1] Written by Karl G. Jöreskog.

In scalar notation, equation (B.1) is

$$y_i = \alpha_i + \beta_{i1}y_1 + \beta_{i2}y_2 + \cdots + \beta_{ii-1}y_{i-1} + \gamma_{i1}x_1 + \gamma_{i2}x_2 + \cdots + \gamma_{iq}x_q + z_i, \tag{B.2}$$

where $i = 1, 2, \ldots, p$ and some of the β's and γ's may be zero. If $\beta_{im} = 0$, y_i does not depend on y_m and if $\gamma_{in} = 0$, y_i does not depend on x_n. The assumptions made are sufficient to show that each equation in (B.2) is a regression equation in the sense that z_i is uncorrelated with all variables appearing on the right side of (B.2). Then the asymptotic variance of the error variance $\text{Var}(z_i) = \hat{\psi}_{ii}$ is given by

$$\text{NAVar}(\hat{\psi}_{ii}) = 2\psi_{ii}^2, \quad i = 1, 2, \ldots, p, \tag{B.3}$$

where ψ_{ii} is the true error variance. An unbiased estimate of the the variance of $\hat{\psi}_{ii}$ is

$$\text{Var}(\hat{\psi}_{ii}) = 2\hat{\psi}_{ii}^2/(N - 1 - s), i = 1, 2, \ldots, p, \tag{B.4}$$

where s is the number of genuine variables appearing on the right side of (B.2).

LISREL estimates model (B.1) by fitting the covariance matrix implied by the model to the sample covariance matrix and the asymptotic variances of parameter estimates are obtained from the information matrix. LISREL does not make use of the fact that each equation can be estimated separately as a regression model. All that LISREL knows is that z is uncorrelated with x. Therefore, LISREL estimates the variance of the error variance using $s = q$ in (B.4), *i.e.*, LISREL uses the aymptotically equivalent formula

$$\text{Est}[\text{Var}(\hat{\psi}_{ii})] = 2\hat{\psi}_{ii}^2/(N - 1 - q), \quad i = 1, 2, \ldots, p. \tag{B.5}$$

A consequence of (B.5) is that the t-value for $\hat{\psi}_{ii}$ will be

$$t(\hat{\psi}_{ii}) = \sqrt{(N - 1 - q)/2}, \quad i = 1, 2, \ldots, p. \tag{B.6}$$

APPENDIX B: WHY ARE T-VALUES EQUAL?

In other words, the t-values of $\hat{\psi}_{ii}$ are the same for all $i = 1, 2, \ldots, p$. This is a paradoxical result, since the t-values are not only the same for all error variances but also for all models containing the same number of x-variables and estimated from the same sample, regardless of what restrictions are imposed on $\mathbf{B}$ and $\mathbf{\Gamma}$. Furthermore, these t-values do not depend on the data other than through the number of x-variables.

This paradox can be observed in the LISREL examples EX44.LS8, EX45A.LS8, and EX44B.LS8 (see Jöreskog & Sörbom, 1996b) and in the SIMPLIS examples EX2A.SPL, EX2B.SPL, and EX3A.SPL (see Jöreskog & Sörbom, 1996c).

The conditions for this result can be slightly extended by allowing for some of the error terms to be correlated. Then (B.3) – (B.5) will hold only for those equations which are regression equations in the sense that the error term z_i is uncorrelated with all variables appearing on the right side of the equation.

Here is an example with $p = 4$ and $q = 3$. This example is based on a fictitious sample covariance matrix assumed to be estimated from 767 cases. The covariance matrix is in the file EQTVAL.COV which is:

```
   1.525
   1.974    7.172
   1.377    4.066    5.438
   1.404    2.602    1.552    3.948
  -0.214   -0.943   -0.590   -0.329    2.736
   0.702    2.310    1.367    1.064   -0.486    3.740
   0.608    2.181    1.533    0.987   -0.355    2.347    3.944
```

A SIMPLIS as well as a LISREL input file for this example follow.

SIMPLIS

The SIMPLIS input is (EQTVAL1.SPL):

```
Illustrating Equal T-Values
Observed Variables: Y1 Y2 Y3 Y4 X1 X2 X3
Covariance Matrix from file EQTVAL.COV
Sample size: 767
Relationships:
```

```
Y1 = X2 X3
Y2 = Y1 X1 X2 X3
Y3 = Y1 Y2 X3
Y4 = Y1 Y2
Number of Decimals = 3
End of Problem
```

This input gives the following results in the output file

$$Y1 = 0.145*X2 + 0.0678*X3, \text{Errorvar.}= 1.382, R^2 = 0.0938$$
$$(0.0278) \quad (0.0271) \quad\quad\quad\quad (0.0707)$$
$$5.222 \quad\quad 2.503 \quad\quad\quad\quad\quad 19.532$$

$$Y2 = 1.063*Y1 - 0.188*X1 + 0.256*X2 + 0.220*X3, \text{Errorvar.}= 3.827, R^2 = 0.463$$
$$(0.0602) \quad (0.0433) \quad (0.0474) \quad (0.0452) \quad\quad\quad\quad (0.196)$$
$$17.639 \quad\quad -4.329 \quad\quad 5.399 \quad\quad 4.866 \quad\quad\quad\quad 19.532$$

$$Y3 = 0.262*Y1 + 0.468*Y2 + 0.0898*X3, \text{Errorvar.}= 3.039, R^2 = 0.440$$
$$(0.0634) \quad (0.0312) \quad (0.0349) \quad\quad\quad\quad (0.156)$$
$$4.131 \quad\quad 14.992 \quad\quad 2.575 \quad\quad\quad\quad\quad 19.532$$

$$Y4 = 0.701*Y1 + 0.170*Y2, \text{Errorvar.}= 2.522, R^2 = 0.360$$
$$(0.0578) \quad (0.0267) \quad\quad\quad\quad (0.129)$$
$$12.128 \quad\quad 6.358 \quad\quad\quad\quad\quad 19.532$$

Note that the t-values for the error variances are all equal to 19.532.

LISREL

The LISREL input (EQTVAL1.LS8) is:

```
Illustrating Equal T-Values
DA NI=7 NO=767
LA=EQTVAL.DAT
CM=EQTVAL.DAT
MO NY=4 NX=3 BE=SD
FI BE(4,3) GA(1,1) GA(3,1) C
   GA(3,2) GA(4,1)-GA(4,3)
OU ND=3
```

This gives the following estimates, standard errors, and t-values for the ψ's.

APPENDIX B: WHY ARE T-VALUES EQUAL?

```
PSI
Note: This matrix is diagonal.

           Y1        Y2        Y3        Y4
        --------  --------  --------  --------
          1.382     3.827     3.039     2.522
         (0.071)   (0.196)   (0.156)   (0.129)
         19.532    19.532    19.532    19.532
```

Now consider the case where z_1 and z_3 are correlated. In the SIMPLIS input (EQTVAL2.SPL) this may be specified by adding the line

```
Let the errors of Y1 and Y3 correlate
```

In the LISREL input (EQTVAL2.LS8), the correlated error may be specified by adding PS=SY on the MO command line and adding the line

```
FR PS(3,1)
```

This gives the following result

```
    PSI

             Y1        Y2        Y3        Y4
          --------  --------  --------  --------
    Y1      1.382
           (0.071)
           19.532

    Y2       - -      3.827
                     (0.196)
                     19.532

    Y3      0.472     - -      3.195
           (0.418)            (0.325)
            1.129              9.825

    Y4       - -      - -       - -      2.522
                                        (0.129)
                                        19.532
```

The t-value for $\hat{\psi}_{33}$ is different from the other t-values because the equation for y_3 is no longer a regression equation.

One can alter the data in any way as long as the sample covariance matrix is positive definite. The t-values will still be 19.532 for all error variances that correspond to regression equations.

C Problems with Analysis of Correlation Matrices

The general rule is that the sample covariance matrix should be analyzed. However, in many behavioral sciences applications, units of measurements in the observed variables have no definite meaning and are often arbitrary or irrelevant. For these reasons, for convenience, and for interpretational purposes, the sample correlation matrix is often analyzed as if it is a covariance matrix. This is a common practice. The usual argument for analyzing a correlation matrix is that one wants a standardized solution. However, this argument is not valid as one can obtain a completely standardized solution even when the covariance matrix is analyzed. The completely standardized solution is obtained by putting SC on the OU command in LISREL syntax or on an Options line in SIMPLIS syntax. A path diagram of the completely standardized solution is obtained by selecting the Standardized Solution when the path diagram is visible.

The analysis of correlation matrices is problematic in several ways. As pointed out by Cudeck (1989), such an analysis may

(a) modify the model being analyzed,
(b) produce incorrect χ^2 and other goodness-of-fit measures, and
(c) give incorrect standard errors.

Problem (a) can occur when the model includes equality constraints or other constrained parameters. For example, if $\lambda_{11}^{(x)}$ and $\lambda_{21}^{(x)}$ are constrained to be equal but the variances $\sigma_{11}^{(xx)}$ and $\sigma_{22}^{(xx)}$ are not equal, then analysis of the correlation matrix will give estimates of

$$\lambda_{11}^{(x)}/\sqrt{\sigma_{11}^{(xx)}} \quad \text{and} \quad \lambda_{21}^{(x)}/\sqrt{\sigma_{22}^{(xx)}} \tag{C.1}$$

which are not equal. Correlation matrices should not be analyzed if the model contains equality constraints of this kind.

Problem (b) can occur in multiple group problems if the observed variables are standardized within groups and there are equality constraints or other constraints across groups.

Problem (c) can occur when a correlation matrix (KM or PM) is analyzed with any method except WLS using the asymptotic covariance matrix of the correlations as estimated by PRELIS. In particular, standard errors will be wrong if a correlation matrix is analyzed with ML.

The main question is whether the standard errors and χ^2 goodness-of-fit measures produced when correlation matrices are used are asymptotically correct. The exact conditions under which this is the case are extremely complicated and give little practical guidance. However, two crucial conditions are

- that the model is scale invariant
- that the fitted covariance matrix has ones in the diagonal

The second condition can be checked by examining the fitted residuals, which should all be zero in the diagonal.

To clarify the issue further, we distinguish between covariance and correlation structures. In principle, all LISREL models are covariance structures, where the variances of Σ as well as the covariances are functions of the parameters. By contrast, in a correlation structure, the diagonal elements of Σ are constants independent of parameters. We now distinguish between four possible cases.

A. A sample covariance matrix is used to estimate a covariance structure.
B. A sample correlation matrix is used to estimate a covariance structure.
C. A sample covariance matrix is used to estimate a correlation structure.
D. A sample correlation matrix is used to estimate a correlation structure.

APPENDIX C: ANALYSIS OF CORRELATION MATRICES

Case A is the standard case in LISREL. In large samples, standard errors and chi-squares will be correct for any method of estimation under multivariate normality of the observed variables and under non-normality if the asymptotic covariance matrix of the sample variances and covariances is provided, see Chapter 4 on *Standard Errors and Chi-squares* in this book.

Case B is a very common situation. Asymptotic variances and covariances of sample correlations are not of the same form as those of the sample variances and covariances, so standard errors will in general be incorrect. However, if the two conditions above hold, chi-squares will still be correct.

For the most common type of models, it is possible to obtain correct standard errors by writing the model in a different way. The following example illustrates this for a confirmatory factor analysis model.

The first example produces wrong standard errors in LISREL syntax (LAWLEY1.LS8):

```
Lawley Factor Analysis Example. Wrong standard errors.
DA NI=9 NO=72 MA=KM
LA
VIS_PERC CUBES LOZENGES PAR_COMP SEN_COMP WRD_MNG
ADDITION CNT_DOT ST_CURVE
KM=LAWLEY.COR
MO NX=9 NK=3 PH=ST
LK
Visual Verbal Speed
FR LX 1 1 LX 2 1 LX 3 1 LX 4 2 LX 5 2 LX 6 2 LX 7 3 LX 8 3 LX 9 1 LX 9 3
OU
```

Next is the command file (LAWLEY2.LS8) that produces the correct standard errors.

```
Lawley Factor Analysis Example. Correct standard errors.
DA NI=9 NO=72 MA=KM
LA
VIS_PERC CUBES LOZENGES PAR_COMP SEN_COMP WRD_MNG
ADDITION CNT_DOT ST_CURVE
KM=LAWLEY.COR
MO NY=9 NE=9 NK=3 LY=DI,FR GA=FI PH=ST PS=DI TE=ZE
LK
Visual Verbal Speed
FR GA 1 1 GA 2 1 GA 3 1 GA 4 2 GA 5 2 GA 6 2 GA 7 3 GA 8 3 GA 9 1 GA 9 3
```

```
CO PS 1 1 = 1 - GA 1 1 ** 2
CO PS 2 2 = 1 - GA 2 1 ** 2
CO PS 3 3 = 1 - GA 3 1 ** 2
CO PS 4 4 = 1 - GA 4 2 ** 2
CO PS 5 5 = 1 - GA 5 2 ** 2
CO PS 6 6 = 1 - GA 6 2 ** 2
CO PS 7 7 = 1 - GA 7 3 ** 2
CO PS 8 8 = 1 - GA 8 3 ** 2
CO PS 9 9 = 1 - GA 9 1**2 - 2*GA 9 1*GA 9 3*PH 3 1 - GA 9 3**2
OU SO
```

The factor loadings with correct standard errors can now be found in the Gamma matrix. It may verified that these two different formulations of the same confirmatory factor analysis model give exactly the same parameter estimates and fit statistics; only the standard errors and t-values differ.

In Case C, the model is defined as a correlation structure $\mathbf{P}(\boldsymbol{\theta})$, with diagonal elements equal to 1 (*i.e., the diagonal elements are not functions of parameters*), it may be formulated as a covariance structure

$$\boldsymbol{\Sigma} = \mathbf{D}_\sigma \mathbf{P}(\boldsymbol{\theta}) \mathbf{D}_\sigma , \qquad (C.2)$$

where $\mathbf{D}_\sigma$ is a diagonal matrix of population standard deviations σ_1, σ_2, ..., σ_k of the observed variables, which are regarded as free parameters. The covariance structure (C.2) has parameters $\sigma_1, \sigma_2, \ldots, \sigma_k, \theta_1, \theta_2, \ldots, \theta_t$. Such a model may be estimated correctly using the sample covariance matrix $\mathbf{S}$. However, the standard deviations $\sigma_1, \sigma_2, \ldots, \sigma_k$, as well as $\boldsymbol{\theta}$ must be estimated from the data and the estimate of σ_i does not necessarily equal the corresponding standard deviation s_i in the sample. When $\mathbf{P}(\boldsymbol{\theta})$ is estimated directly from the sample correlation matrix $\mathbf{R}$, standard errors and χ^2 goodness-of-fit values will in general not be correct.

Consider Case D. To obtain correct asymptotic standard errors in LISREL for a correlation structure when the correlation matrix is analyzed, the WLS method must be used with a weight matrix $\mathbf{W}^{-1}$, where $\mathbf{W}$ is a consistent estimate of the asymptotic covariance matrix of *the correlations being analyzed*. Such a $\mathbf{W}$ may be obtained with PRELIS under non-normal theory. PRELIS can estimate such a $\mathbf{W}$ also for a correlation matrix containing polychoric and/or polyserial correlations.

APPENDIX C: ANALYSIS OF CORRELATION MATRICES

The asymptotic covariance matrix $\mathbf{W}$ produced by PRELIS is a consistent estimate of the covariance matrix

$$\mathbf{r} = (r_{21}, r_{31}, r_{32}, r_{41}, r_{42}, \ldots)$$

The diagonal elements of the correlation matrix are not included in this vector. The number of distinct elements in $\mathbf{W}$ are

$$\frac{1}{2}k(k-1)\left[\frac{1}{2}k(k-1)+1\right],$$

where $k = p + q$ is the number of observed variables in the model. In fitting a correlation structure $\mathbf{P}(\boldsymbol{\theta})$ to a correlation matrix using WLS, LISREL minimizes the fit function

$$F(\boldsymbol{\theta}) = (\mathbf{r} - \boldsymbol{\rho})' \mathbf{W}^{-1} (\mathbf{r} - \boldsymbol{\rho}) \tag{C.3}$$

where

$$\boldsymbol{\rho}' = (\rho_{21}(\boldsymbol{\theta}), \rho_{31}(\boldsymbol{\theta}), \rho_{32}(\boldsymbol{\theta}), \rho_{41}(\boldsymbol{\theta}), \ldots, \rho_{k,k-1}(\boldsymbol{\theta})).$$

This approach assumes that the diagonal elements of $\mathbf{P}(\boldsymbol{\theta})$ are fixed ones and not functions of parameters.

WLS may also be used to fit ordinary LISREL models (*i.e., covariance structures*) to sample correlation matrices (Case B). This is especially useful when polychoric and polyserial correlations are analyzed. A small problem arises here because the fit function (C.3) is not a function of the diagonal elements of $\mathbf{P}(\boldsymbol{\theta})$, and, as a consequence, parameters such as the diagonal elements of $\boldsymbol{\Theta}_\epsilon$ and $\boldsymbol{\Theta}_\delta$ cannot be estimated directly. However, they can of course be estimated afterwards.

A better and more general approach is to add the term

$$\sum [1 - \sigma_{ii}(\boldsymbol{\theta})]^2 \tag{C.4}$$

to the fit function (C.3). The advantages of this approach are

- Estimates of all parameters can be obtained directly even when constraints are imposed
- When the diagonal elements of Θ_ϵ and Θ_δ are free parameters, the WLS solution will satisfy

$$\text{Diag}(\hat{\Sigma}) = \mathbf{I}.$$

This approach can be generalized further by replacing the one in (C.4) with a variance s_{ii} which has been estimated or obtained separately from the correlations. Such variances can be obtained with PRELIS for ordinal variables for which the thresholds are assumed to be equal.

D PRELIS Syntax Overview

This section provides a convenient reference to PRELIS 2. It has been updated with the syntax introduced with the new statistical features in Chapter 3.

The commands are arranged in logical order. The diagram is constructed according to the following conventions:

- Maximum line length is 127 columns. Commands may be continued over several lines by adding a space followed by a C (for 'continue') on the current line. A keyword and its specified value should appear on the same line: start a keyword on a new line if its specified value would extend past column 127.

- Square brackets [ ] enclose optional specifications. The brackets themselves should not be coded.

- Boxes enclose alternative specifications. Only one element of the list may be entered. A **boldface** element indicates the default specification.

- Parentheses ( ) must be entered exactly as shown.

- Equals signs = are required.

- Uppercase elements are commands, keywords, keyword values, or options. They must be entered as they appear, or they may be lengthened (LABELS instead of LA, for example).

- Lowercase elements describe information to be filled in by the user.

- Use blanks to separate command names, keywords, and options.

- An exclamation mark (!) or the slash-asterisk combination (/*) may be used to indicate that everything that follows on this line is to be regarded as comments. Blank (empty) lines are accepted without the ! or /*.
- Command order is important. After optional title lines, the DA command should appear first, the OU command should be last. The LA command should be placed before any other command using named variables (instead of variable numbers). The RA command may appear anywhere.

  All other commands may be used more than once. Below, they are given in logical order. Note that PRELIS 2 processes recoding and transformation first, then selection, finally missing values. In exceptional cases, this may necessitate more than one run, each time saving the transformed raw data.
- For format statements, see the *PRELIS 2: User's Reference Guide*. No format statement means free format: the data are separated by a space, comma, and/or return character.
- An *italic* element indicates a new feature of PRELIS 2.

D.1 PRELIS syntax diagram

["title line"]

[...]

D.1.1 Data input commands

[SY=[filename].PSF]   See page 168

DA NI=k[,l,m,...] [NO= | 0 / number of cases |] [TR= | LI / PA |]

[MI=global missing value(s)] [RP=no. of repetitions]

[LA [[FI]=filename [FO]]]

[(character variable format statement)]

[variable labels]

RA [FI]=filename[,filename,filename,...] [FO] [RE]

[(variable format statement)
  [(variable format statement)]
  [...]]

D.1.2 Data manipulation commands

Scale types

[FI varlist]

[CA | varlist / ALL |]

[CB | varlist / ALL |]

[CE | varlist / ALL |]

[CO | varlist / ALL |]

[OR | varlist / ALL |]

Recode and label categories

[RE | varlist / ALL | OLD=valuerange[,valuerange, . . .] NEW=value[,value, . . .]]

[*CL* | varlist / ALL | n_1=clab$_1$ n_2=clab$_2$ . . .]

Transformation and creation of variables

[*NE* newvar=function of old variables, *NRAND* and/or *URAND*]

[LO | varlist / ALL | [AL= | 0 / α-value |] [BE= | 1 / β-value |]]

[PO | varlist / ALL | [AL= | 0 / α-value |] [BE= | 1 / β-value |] [GA= | 1 / γ-value |]]

Select cases and variables

[SC varlist | = value / [> value] [< value] |]

[*SC* [CASE= | ODD / EVEN |] [> number] [< number]]

D.1 PRELIS SYNTAX DIAGRAM

[SD varlist [= value   ]]
            [> value] [< value]

[*SE* varlist]

D.1.3 Treatment of missing values

[*IM* (lvarlist) (Mvarlist) [VR= .5   ] [XN] [XL]]
                                 value

[MI valuerange[,valuerange,...] varlist]

D.1.4 Analysis and output commands

[WE variable]

[*HT* varlist]

[*ET* varlist]

⎡ *FT*=filename varlist1 ⎤
⎢ [*FT* varlist2] ALL   ⎥
⎣ [...]   ⎦

[*MT* varlist]
       ALL

[*RG* y-varlist   ON x-varlist   [*WITH* z-varlist] [*RES*=newvar]]
      ALL   ALL   ALL

[*EQ* y-varlist = x-varlist [*WITH* z-varlist]]
      ALL   ALL   ALL

[*FA* [NF=number] [FS]]

[*PC* [NC=number] [PS]]

[*NS* varlist]

[*FS* filename.MSF]

OU [MA= AM] [RA=filename] [AC=filename]]
        CM   SR=filename   SA=filename
        KM
        MM
        OM
        PM
        RM
        TM

[WI=format width] [ND=no. of decimals]

[AM=filename] [CM=filename] [KM=filename] [MM=filename]

[OM=filename] [PM=filename] [*RM*=filename] [*TM*=filename]

[SM=filename] [SV=filename] [*TH*=filename]

[*BM*=filename] [*ME*=filename] [*SD*=filename]

[*YE*=filename] [*YS*=filename]

[*BS*=no. of bootstrap samples] [*SF*=sample fraction]

[*IX*=integer starting value for the random number generator]

[PA] [PV] [WP] [XB] [XT] [*XM*] [*XO*[=number]]

E LISREL Syntax Overview

This appendix provides a convenient reference to LISREL 8. It has been updated with the syntax introduced with the new statistical features in Chapter 3.

The commands are arranged in logical order. The diagram is constructed according to the following conventions:

- Square brackets [] enclose optional specifications. The square brackets themselves should not be coded.
- Boxes enclose alternative specifications. Only one element of the list may be entered. A **boldface** element indicates the default specification.
- Parentheses () must be entered exactly as shown.
- Equals signs = are required.
- Uppercase elements are commands, keywords, keyword values, or options. They must be entered as they appear, or they may be lengthened (LABELS instead of LA, for example). Thus, except for the ALL option on the VA and ST commands and the PATH DIAGRAM command (though PD may be used instead), everything has *two* significant characters.
- Lowercase elements describe information to be filled in by the user.
- Use blanks to separate command names, keywords, and options.
- An exclamation mark (!) or the slash-asterisk combination (/*) may be used to indicate that everything that follows on this line is to be regarded as comments. Blank (empty) lines are accepted without the ! or /*.

- Command order is important, see the section *Order of commands* in the *LISREL 8: User's Reference Guide*
- For format statements, see the section *FORTRAN format statements in the command file* in the *LISREL 8: User's Reference Guide*. No format statement means free format: the data are separated by a space, comma, and/or return character.
- A parameter matrix element should be written as a parameter matrix name (LY, LX, BE, GA, PH, PS, TE, TD, TH, TY, TX, AL, or KA), followed by row and column indexes (or linear indexes) of the specific element. Row and column indexes may be separated by a comma and enclosed in parentheses, like LY(3,2), LX(4,1), or separated from the matrix name and each other by spaces, like LY 3 2 LX 4 1.
- The order of the form and mode values for the parameter matrices on the MO command line is optional, but if both are given, a comma in between is required.
- An *italic* element indicates a new feature of LISREL 8

The maximum line length in a command file is 127 columns. Commands may be continued over several lines by adding a space followed by a C (for 'continue') on the current line. A keyword and its specified value should appear on the same line: start a keyword on a new line if its specified value would extend past column 127. Note that this requirement has not been followed in the syntax diagram below.

E.1 LISREL syntax diagram

["title line"]

[...]

E.1.1 Input specification commands

[SY=[filename].DSF]   See page 169

[PS=[filename].PSF]   See page 171

DA NI=k NO=number of cases [NG= | 1 / number of groups |] [MA= | **CM** / AM / KM / MM / OM / PM / *RM* / *TM* |]

[XM=global missing value] [*RP*=no. of repetitions]

[LA [[FI]=filename [FO] [RE]]]

[(character variable format statement)]

[y and x labels]

[RA [[FI]=filename [FO] [RE]]]

[(variable format statement)]

[data records]

[| CM / KM / MM / OM / PM / *RM* / *TM* | [[FI]=filename [FO] [RE] [| **SY** / FU |]]]

[(variable format statement)]

[data records]

[ME [[FI]=filename [FO] [RE]]]

[(variable format statement)]

[data records]

[SD [[FI]=filename [FO] [RE]]]

[(variable format statement)]

[data records]

[$\begin{array}{|c|}\text{AC} \\ \text{WM}\end{array}$ [FI]=filename]

[$\begin{array}{|c|}\text{AV} \\ \text{DM}\end{array}$ [FI]=filename]

[SE [[FI]=filename]]

[variable names]

E.1.2 General analysis specification commands

[*RG* [list of y-variables] ON [list of x-variables] [WITH [list of z-variables]]

[*FA* [NF=[number]]]

[*PC* [NC=[number]]]

E.1 LISREL SYNTAX DIAGRAM

E.1.3 LISREL model specification commands

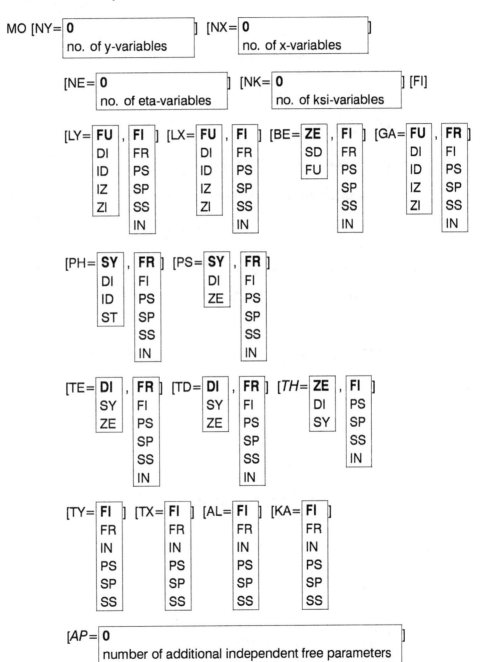

[LK [[FI]=filename [FO] [RE]]]

[(character variable format statement)]

[ksi labels]

[LE [[FI]=filename [FO] [RE]]]

[(character variable format statement)]

[eta labels]

[FR list of parameter matrix elements]

[FI list of parameter matrix elements]

[EQ list of parameter matrix elements]

[*CO* parameter matrix element=expression with other parameters]

[*IR* list of parameter matrix elements [>number] [<number]]

[PA [[FI]=filename [FO] [RE]] matrix name]

[(integer format statement)]

[pattern records]

[VA numerical value | list of parameter matrix elements
                    | ALL]

[ST numerical value | list of parameter matrix elements
                    | ALL]

[MA [[FI]=filename [FO] [RE]] matrix name]

[(variable format statement)]

[records of matrix values]

[PL list of parameter matrix elements [FROM a TO b]]

[NF list of parameter matrix elements]

E.1.4 Output specification commands

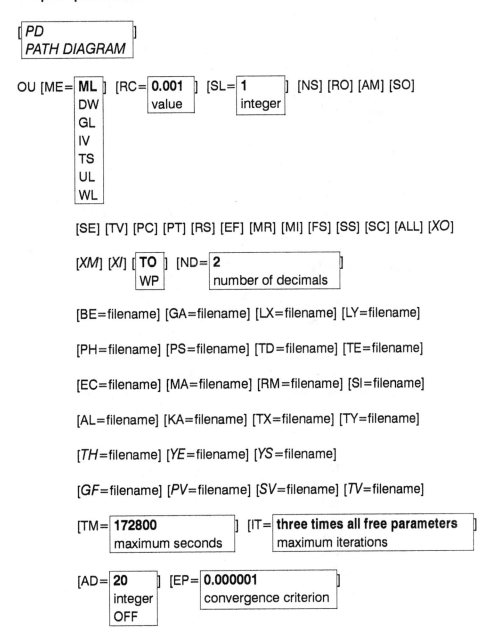

E.2 Notation

The Greek Alphabet

α	A	alpha
β	B	beta
γ	Γ	gamma
δ	Δ	delta
ϵ	E	epsilon
ζ	Z	zeta
η	H	eta
θ	Θ	theta
ι	I	iota
κ	K	kappa
λ	Λ	lambda
μ	M	mu
ν	N	nu
ξ	Ξ	xi, ksi
o	O	omicron
π	Π	pi
ρ	P	rho
σ	Σ	sigma
τ	T	tau
υ	Υ	upsilon
ϕ	Φ	phi
χ	X	chi
ψ	Ψ	psi
ω	Ω	omega

Typical LISREL Notation

x, y	observed variables
ξ, η	latent variables
ζ, δ, ϵ	error variables
Λ_x, Λ_y	factor loadings
B, Γ	structural parameters
$\alpha, \kappa, \tau_x, \tau_y$	mean parameters
Φ, Ψ	covariance matrices
$\Theta_\delta, \Theta_\epsilon, \Theta_{\delta\epsilon}$	error covariance matrices
$\hat{\Lambda}_x$	estimate of Λ_x

Other Notation

$\mathbf{x}$	column vector
$\mathbf{x}'$	row vector
$\mathbf{X}$	matrix
$\mathbf{X}'$	matrix transpose
$\mathbf{X}^{-1}$	matrix inverse
$[x_{ij}]$	matrix element
$\|\mathbf{X}\|$	determinant of a square matrix $\mathbf{X}$
$\text{tr}(\mathbf{X})$	trace of $\mathbf{X}$ (sum of diagonal elements of a square matrix)
Greek letters	population parameters, latent random variables
Roman letters	observed random variables

F Multilevel Syntax Overview

This appendix provides a convenient reference to the multilevel syntax as described in Chapter 2.

The commands are arranged in logical order, with the exception of the OPTIONS command that has to come first. The diagram is constructed according to the following conventions:

- Square brackets [] enclose optional specifications. The square brackets themselves should not be coded.
- Boxes enclose alternative specifications. Only one element of the list may be entered. A **boldface** element indicates the default specification.
- Equals signs = are required.
- A semicolon ; signals the end of a command.
- Uppercase elements are commands, keywords, keyword values, or options. They must be entered as they appear, or they may be lengthened (TOEPLITZ instead of TOEP, for example).
- Lowercase elements describe information to be filled in by the user.
- Some command names contain a lowercase 'n.' This should be replaced with an integer (1, 2, 3, ...) indicating the level of the hierarchy. For example, enter ID1 or ID2, *not* IDn.
- Two ID commands are required for a level-2 model, three ID commands (ID1, ID2, ID3) for a level-3 model. ID1 may be omitted with a multivariate model or a model with no random components on level 1.
- Use blanks to separate command names, keywords, and options.

- Blank (empty) lines are accepted within the input file.
- Command order is unimportant, with the exception that the OPTIONS command always comes first.
- The VAR3:VAR10 notation may be used to indicate a range of consecutively numbered variables, in this case VAR3, VAR4, ..., VAR10.
- The format for entering a covariate on the FIXED command is: var1 var 2 ... var$n$ covariate*var1 covariate*2 ... covariate*var$n$, for example, with the covariate being the variable gender: constant time timesq gender gender*time gender*timesq.

The maximum line length in a command file is 80 columns. Commands may be continued over several lines. A keyword and its specified value should appear on the same line: start a keyword on a new line if its specified value would extend past column 80.

F.1 Multilevel syntax diagram

OPTIONS [OLS= | YES　 |] [CONVERGE= | 0.001 |] [MAXITER= | 10 |]
　　　　　　　| NONE |　　　　　　　　| numerical value |　　　　　| integer |

　　　　[OUTPUT= | STANDARD |] [LINK= | 0 |] [ADD= real numberf between 0
　　　　　　　　 | BAYES | | 1 |
　　　　　　　　 | RESIDUAL |
　　　　　　　　 | ALL |
　　and 1] ;

[TITLE = descriptive title for the specific analysis ;]

IDn = name of variable ;

RESPONSE = list of response variables ;

FIXED = names of variables included as fixed effects in the model ;

RANDOMn = names of variables random on level n of the model ;

[COVnPAT = | DIAG | ;]
　　　　　　| TOEP |
　　　　　　| INTR |
　　　　　　| MA1 |
　　　　　　| LOGIT |
　　　　　　| user specified |

[COVnVAL = starting values for level n random coefficient covariance matrix ;]

[FIXVAL = starting values for fixed effect parameters ;]

[CONTRAST = name of contrast file and directory path as needed ;]

[MISSING_DAT = integer value ;]

[MISSING_DEP = integer value ;]

[SUBPOP = names of variables to be used to construct subpopulations ;]

APPENDIX F: MULTILEVEL SYNTAX OVERVIEW

References

Anderson, T.W., & Rubin, H. (1956)
   Statistical inference in factor analysis.
   In *Proceedings of the Third Berkeley Symposium*, Volume V. Berkeley: University of California Press.

Bollen, K. A. (1995)
   Structural equation models that are nonlinear in latent variables: A least squares estimator.
   In P.M. Marsden (Ed.): *Sociological Methodology 1995*. Cambridge, MA: Blackwell.

Bollen, K.A. (1996)
   An alternative two stage least squares (2SLS) estimator for latent variable equations.
   Psychometrika, **61**, 109–121

Bollen, K.A. & Paxton, P. (1998)
   Two-stage least squares estimation of interaction effects.
   Pp. 125–151 in G.A. Marcoulides & R.E. Schumacker (Eds): *Interaction and nonlinear effects in structural equation models*. Mahwah, N.J: Lawrence Erlbaum Associates, Publishers.

Browne, M.W. (1974)
   Generalized least squares estimators in the analysis of covariance structures.
   South African Statistical Journal, **8**, 1–24.

Browne, M.W. (1977)
   Generalized least squares estimators in the analysis of covariance structures.
   Pp. 205–226 in D.J. Aigner and A.S. Goldberger (Eds.): *Latent variables in socio-economic models*. Amsterdam: North-Holland.

Browne, M.W. (1984)
: Asymptotically distribution-free methods for the analysis of covariance structures.
: *British Journal of Mathematical and Statistical Psychology*, **37**, 62-83.

Browne, M.W. (1987)
: Robustness of statistical inference in factor analysis and related models.
: *Biometrika*, **74**, 375–384.

Browne, M. W. & Cudeck, R. (1993)
: Alternative ways of assessing model fit.
: In K. A. Bollen & J. S. Long (Editors): *Testing Structural Equation Models*. Sage Publications.

Bryk, A.S. & Raudenbush, S.W. (1992)
: Hierarchical Linear Models.
: Sage.

Cudeck, R. (1989)
: Analysis of correlation matrices using covariance structure models.
: *Psychological Bulletin*, **105**, 317–327.

Finn, J.D. (1974)
: *A general model for multivariate analysis*.
: New York: Holt, Reinhart and Winston.

French, J.V. (1951)
: The description of aptitude and achievement tests in terms of rotated factors.
: *Psychometric Monographs*, **5**.

Goldberger, A.S. (1964)
: *Econometric theory*.
: New York: Wiley.

Goldstein, H. (1987)
: Multilevel models in educational and social research.
: C. Griffin and Co., London, Oxford University Press, New York.

Goldstein, H. (1995)
: Multilevel Statistical Models.
: Edward Arnold.

REFERENCES

Guilford, J.P. (1956)
> The structure of intellect.
> *Psychological Bulletin*, **53**, 267–293.

Hendrickson, A.E. & White, P.O. (1964)
> Promax: A quick method for rotation to oblique simple structure.
> *British Journal of Mathematical and Statistical Psychology*, **17**, 65-70.

Holzinger, K., & Swineford, F. (1939)
> *A study in factor analysis: The stability of a bifactor solution.*
> Supplementary Educational Monograph no. 48.
> Chicago: University of Chicago Press.

Hu, L., Bentler, P.M., & Kano, Y. (1992)
> Can test statistics in covariance structure analysis be trusted?
> *Psychological Bulletin*, Vol. **112**, 351–362.

Hägglund, G. (1982)
> Factor analysis by instrumental variable methods.
> *Psychometrika*, **47**, 209–222.

Jöreskog, K.G. (1967)
> Some contributions to maximum likelihood factor analysis.
> *Psychometrika*, **32**, 443–482.

Jöreskog, K.G. (1969)
> A general approach to confirmatory maximum likelihood factor analysis.
> *Psychometrika*, **34**, 183–202.

Jöreskog, K.G. (1977)
> Factor analysis by least-squares and maximum-likelihood methods.
> Pp. 125–153 in K. Enslein, A. Ralston, & H.S. Wilf (Eds.):
> *Statistical methods for digital computers*. New York: Wiley.

Jöreskog, K.G. (1979)
> Basic ideas of factor and component analysis.
> In K.G. Jöreskog & D. Sörbom: *Advances in factor analysis and structural equation models*. Cambridge, Mass.: Abt Books, 5–20.

Jöreskog, K.G. (1981)
> Analysis of covariance structures.
> *Scandinavian Journal of Statistics*, **8**, 65–92.

Jöreskog, K.G. (1990)
New developments in LISREL: Analysis of ordinal variables using polychoric correlations and weighted least squares.
Quality and Quantity, **24**, 387–404.

Jöreskog, K.G. (1993)
Testing structural equation models.
In K.A. Bollen & J.S. Long (Eds), *Testing Structural Equation Models*.
Sage Publications.

Jöreskog, K.G. (1994).
On the estimation of polychoric correlations and their asymptotic covariance matrix.
Psychometrika, **59**, 381–389.

Jöreskog, K.G., & Sörbom, D. (1996a)
PRELIS 2 *User's Reference Guide*.
Chicago: Scientific Software International.

Jöreskog, K.G., & Sörbom, D. (1996b)
LISREL 8 *User's Reference Guide*.
Chicago: Scientific Software International.

Jöreskog, K.G., & Sörbom, D. (1996c)
LISREL 8: *Structural Equation Modeling with the SIMPLIS Command Language*.
Chicago: Scientific Software International.

Jöreskog, K.G., & Yang, F. (1996).
Nonlinear structural equation models: The Kenny–Judd model with interaction effects.
Pp. 57–88 in G.A. Marcoulides & R.E. Schumacker (Eds):
Advanced structural equation modeling: Issues and techniques.
Lawrence Erlbaum Associates, Publishers.

Kaiser, H.F. (1958)
The varimax criterion for analytical rotation in factor analysis.
Psychometrika, **23**, 187–200.

Kanfer, R. & Ackerman, P. L. (1989)
Motivation and cognitive abilities: an integrative/aptitude-treatment interaction approach to skill acquisition.
Journal of Applied Psychology, Monograph, **74**, 657–690.

REFERENCES

Kenny, D.A., & Judd, C.M. (1984).
Estimating the nonlinear and interactive effects of latent variables.
Psychological Bulletin, **96**, 201–210.

Klein, L.R. (1950)
Economic fluctuations in the United States 1921–1941.
Cowles Commission Monograph No. 11. New York: Wiley.

Kreft, I. & de Leeuw, J. (1998)
Introducing multilevel modeling.
Sage.

Lawley, D.N., & Maxwell, A.E. (1971)
Factor analysis as a statistical method, (2nd edition).
London: Butterworths.

Longford, N.T. (1987)
A fast scoring algorithm for maximum likelihood estimation in unbalanced mixed models with nested effects.
Biometrika, **741**, 4, 817–827.

Magnus, J.R. & Neudecker, H. (1988)
Matrix differential calculus with applications in statistics and econometrics.
New York, Wiley.

Mardia, K.V., Kent, J.T., & Bibby, J.M. (1980)
Multivariate analysis.
New York: Academic Press.

Mortimore, P., Sammons, P., Stoll, L., Lewis, D. & Ecob, R. (1988)
School Matters, the Junior Years.
Wells, Open Books.

Prosser, R., Rasbash, J. & Goldstein, H. (1991)
ML3 software for Three-level Analysis, Users' Guide for V.2.
Institute of Education, University of London.

Rasbash, J. (1993)
ML3E Version 2.3 Manual Supplement.
University of London: Institute of Education..

Reyment, R. & Jöreskog K.G. (1993)
Applied factor analysis in the natural sciences.
Cambridge: Cambridge University Press.

Satorra, A. (1987)
> Alternative test criteria in covariance structure analysis: A unified approach.
> *Psychometrika*, **54**, 131–151.

Satorra, A. (1993)
> Multi-sample analysis of moment structures: Asymptotic validity of inferences based on second-order moments.
> Pp. 283–298 in K. Haagen, D.J. Bartholomew, & M. Deistler (Editors): Statistical Modelling and Latent Variables. Elsevier Science Publishers.

Satorra, A., & Bentler, P.M. (1988)
> Scaling corrections for chi-square statistics in covariance structure analysis.
> Proceedings of the Business and Economic Statistics Section of the American Statistical Association, 1988, 308–313.

Schumacker, R.E. & Marcoulides, G.A. (Eds) (1998)
> *Interaction and nonlinear effects in structural equation models.*
> Mahwah, N.J: Lawrence Erlbaum Associates, Publishers.

Sörbom, D. (1974)
> A general method for studying differences in factor means and factor structures between groups.
> British Journal of Mathematical and Statistical Psychology, Vol. **27** 229–239.

Steiger, J.H. (1990)
> Structural model evaluation and modification: An interval estimation approach.
> *Multivariate Behavioral Research*, **25**, 173–180.

Theil, H. (1971)
> *Principles of econometrics.*
> New York: Wiley.

Thurstone, L.L. (1938)
> Primary mental abilities.
> *PsychometricMonographs*, **1**.

Tintner, G. (1952)
> *Econometrics.*
> New York: Wiley.

REFERENCES

Yuan, K-H., & Bentler, P.M. (1997)
   Mean and covariance structure analysis: Theoretical and practical improvements.
   Journal of the American Statistical Association, Vol. **92**, 767–774.

REFERENCES

Author Index

Ackerman, 92
Agresti, 32
Anderson, 31, 156

Bartlett, 156
Bentler, 180
Bibby, 157, 158
Bishop, 32
Bollen, 172, 173
Browne, 25, 26, 147, 155, 180, 191, 196
Bryk, 14, 72

Cramer, 34
Cudeck, 152, 155, 209

du Toit, 27, 63, 71

Fienberg, 32
Finn, 134
French, 144

Goldberger, 131, 136, 138, 142
Goldstein, 26, 86
Guilford, 144

Hägglund, 147

Hendrickson, 151
Holland, 32
Holzinger, 147, 183
Hu, 181

Jöreskog, 129, 144, 146, 147, 151, 154, 155, 157, 168, 172, 175, 182, 190, 191, 203, 205
Judd, 172

Kaiser, 151
Kanfer, 92
Kano, 181
Kenny, 172
Kent, 157, 158
Klein, 137

Lawley, 154
Longford, 14
Magnus, 193

Malinvaud, 28
Marcoulides, 171
Mardia, 157, 158
Maxwell, 154
McCulloch, 25
Morrison, 30

Mortimer, 102
Neudecker, 193

Paxton, 172, 173
Potthoff, 15

Rasbash, 14
Raudenbush, 14, 72
Reyment, 157
Roy, 15
Rubin, 156

Sörbom, 147, 154, 155, 168, 175, 182, 189, 190, 203, 205
Satorra, 180, 191, 199
Schumacker, 171
Steiger, 154
Swineford, 147, 183

Theil, 131, 138
Thurstone, 144
Tintner, 141

White, 151

Yang, 172
Yuan, 181

Subject Index

ADD (keyword), 45
ADF, 3
Air traffic control data, 92
Asymptotic (co)variance, 3, 7
Asymptotic chi-square, 181
Augmented moment matrix, 8, 198, 202

Blank line, 42
Bootstrap, 4, 182

C1, 180
C2, 180
C3, 180
C4, 180
Chi-square, 179
Cluster design, 14
Collinearity, 14
Command language, 10
Command order, 216
Completely standardized solution, 209
Complex survey, 14
Complex variation, 89
Confirmatory analysis, 143
Confirmatory factor analysis, 145
CONS.PR2, 125
CONS.PSF, 125
Constraint, 35, 48, 50

Contrast, 28
CONTRAST (command), 57
CONVERGENCE (keyword), 43
Correlation matrix, analysis of, 209
COVnPAT (command), 50
COVnVAL (command), 55
COVnVAL (command), example, 56
Covariance structure, 192, 199
Covariate, adding a, 87

Data, 2
Data file, 40
Data system file, 169
Decision table, 155
DSF file, 169
DWLS, 3, 6

Econometric model, 136
EDUC.PR2, 121
EDUC.PSF, 121
Eigenvalue, 160
Eigenvector, 160
Empirical Bayes estimate, 30, 44, 69, 84
EQTVAL1.LS8, 206
EQTVAL1.SPL, 205
EQTVAL2.LS8, 207

EQTVAL2.SPL, 207
Error in measurement, 5
Estimation, 6
EX2.PR2, 40
Example, repeated measures, 70
Example: Exploratory Factor Analysis, 147
Example: Normalizing variables, 162, 165
Example: Prediction of Grade Averages, 134
Example: Principal components, 158
Example: The Kenny–Judd Model, 172
Example: Tintner's Meat Market Model, 141
Exploratory analysis, 143
Exploratory factor analysis, 143, 144

FA (command), 149
Factor Analysis (command), 148
Factor analysis,
    basic idea, 144
    exploratory, 143
Factor loading, 151
Factor score, 155
    merge, 156
    usage, 156
Fit function, 7, 194, 197, 199
FIXED (command), 49
    example, 50
Fixed effect, 99
FIXVAL (command), 56
Format statement, 216, 222
Frequency table, 45

FS (keyword), 155
Fully unconditional model, 72

General covariance structure, 10
Generalized least squares estimator, 24
GF file, 182
GLS, 6
GRAV.PR2, 134
GRAV.RAW, 134
Greek alphabet, 228
Growth curve modeling, 13

ID (command), 46
    example, 47
IGLS estimator, 27
INCOME.PR2, 117
INCOME.PSF, 116
Input file, 39
Interaction effect, 171
Interactive LISREL, 12
Iterative generalized least squares estimator, 27
IV, 6

JSP.PSF, 102
JSP1.OUT, 105
JSP1.PR2, 105
JSP2.PR2, 109

KANFER.PSF, 92
KANFER1.PR2, 93
KANFER2.PR2, 97
KANFER3.PR2, 99
Kenny–Judd model, 172
KJSIM.PR2, 176

KJTSLS1.PR2, 173
KJTSLS2.PR2, 173
Klein's model, 137
KLEIN.RAW, 138
KLEIN1.PR2, 138
KLEIN2.PR2, 140
KLEIN3.PR2, 140
Kurtosis, 164

Latent curve modeling, 13
Latent variable, 6
Latent variable score, 171
LAWLEY1.LS8, 211
LAWLEY2.LS8, 211
Level-2 model, 17
Level-3 logit model, 32
Level-3 model, 19
    example, 116
Likelihood ratio test, 31
Line continuation, 215, 222
Line length, 215, 222, 230
Linear growth, 76
LINK (keyword), 45
LISREL
    model, 4
    notation, 228
    syntax conventions, 221
    syntax diagram, 223
Logit model, 32

Maximum likelihood, 7
Maximum likelihood estimator, 28
MAXITER (keyword), 43
Measurement error, 5, 145
MISSING_DAT (command), 59
MISSING_DEP (command), 60

ML, 6, 7
Model system file, 170
Moment matrix, 157
Monte Carlo, 4, 182
MOUSE.PSF, 71
MOUSE1.PR2, 73
MOUSE2.PR2, 76
MOUSE3.OUT, 78
MOUSE3.PR2, 78
MOUSE4.BA2, 83
MOUSE4.PR2, 81
MOUSE4.RES, 85
MOUSE5.PR2, 87
MOUSE6.PR2, 89
MSF file, 170
Multi-stage sample design, 14
Multigroup analysis, 9
Multilevel logit model, 32
Multilevel syntax
    conventions, 229
    diagram, 231
Multiple group, 199
Multitrait-multimethod, 145
Multivariate normality, 164

NCP, 182
Non-linear growth, 81
Non-linear model, 96, 171
Non-normal variable, 165
Normal score, 161
Normal score, formula, 162
Normalized variable, 166
NPV2.SPL, 149, 165
NPV3.LS8, 152
NPV4.SPL, 152, 165
NPV5.LS8, 152
NPV6.LS8, 152

NPV7A.PR2, 152
NPV7B.PR2, 153
NPV7C.PR2, 156
NPV7D.PR2, 156
NPVNSC1.PR2, 162
NS (command), 163
Number of factors, 153

Oblique solution, 151
OLS, 130
OLS (keyword), 42
OLS formula, 132
OPTIONS (command), 42
    example, 46
Order condition, 136, 138
Order of commands, 216
Ordinal variable, 2
Ordinary least squares estimator, 16
Ordinary least-squares, 130
Orthogonal solution, 151
Output, 12
OUTPUT (keyword), 43
Output file, 63

Panel design, 145
Parameter estimation, 22
Parameters, 6
Path diagram, 11
PC (command), 157
PCEX1.LS8, 158
PCEX3.PR2, 161
PRELIS, 1
    syntax conventions, 215
    syntax diagram, 217
    system file, 40, 168
Principal component, 157

Principal Components (command), 158
Probit regression, 3
Promax, 149, 151
PS (option), 157
PS command, 171
PSF file, 40, 168
PSFfile, 171

RANDOM (command), 47
Random coefficient modeling, 13
Rank condition, 136
Reference variable, 146
Regress (command), syntax, 133
Regression model, 134
Repeated measurement, 70
Required command, 41
Residual, 44, 69, 85
Residual analysis, 133
RESPONSE (command), 48
    example, 49
Restricted solution, 146
RG (command), 132
    syntax, 133
RMSEA, 154, 182
Robust chi-square, 191
Robust standard error, 191

Sample, 161
sample size, 10
Satorra-Bentler chi-square, 180
Saturated model, 193
SC (option), 152, 209
Scale type, 2
SE (command), 153
Semi-colon, in syntax, 42
Simulation, 4, 175

Simulation study, 182
Single group, 192
Skewness, 164
Specific factor, 144
SPV.PR2, 184
SPV.RAW, 183
SPV1.SPL, 186
SPV3.SPL, 189
Standard error, 28, 179
Standardized solution, 147, 209
Statistical inference, 27
SUBPOP (command), 61
SY (command), 40, 169
Syntax file, 41
Syntax overview, 41
Syntax rules, 42
System file, 168
System File (command), 169

t-value, 203
Test of model, 6, 9
TINTNER1.SPL, 142
TINTNER2.LS8, 143
TITLE (command), 62
Toeplitz matrix, 51
TSLS, 6, 130
    estimate, properties, 175
    estimator, 131
    solution, 151
    standard error, 151
Two-stage least-squares, 130
Type of matrix, 2

ULS, 6
Unrestricted solution, 146
Unrotated solution, 151

Variable, 2
    non-normal, 165
    normalized, 166
    ordinal, 2
    reference, 146
Variance decomposition, 72, 93, 104
Varimax, 149, 151

Weight matrix, 194
WEIGHT1 (command), 62
Weighted least squares, 7
Windows interface, 11
WLS, 3, 6, 7
WMCD1.PR2, 166
WMSD2.PR2, 167

XI (keyword), 183
XO (keyword), 177